Restoring Paradise

Restoring Paradise

RETHINKING AND REBUILDING NATURE IN HAWAI'I

Robert J. Cabin

A Latitude 20 Book
UNIVERSITY *of* HAWAI'I PRESS
HONOLULU

© 2013 University of Hawaiʻi Press
All rights reserved
Printed in the United States of America

18 17 16 15 14 13   6 5 4 3 2 1

Library of Congress Cataloging-in-Publication Data
Cabin, Robert J., author.
  Restoring paradise : rethinking and rebuilding nature in Hawaiʻi / Robert J. Cabin.
    pages cm
  "A latitude 20 book."
  Includes bibliographical references and index.
  ISBN 978-0-8248-3693-1 (pbk. : alk. paper)
  1. Restoration ecology—Hawaii.
  2. Nature conservation—Hawaii.
  I. Title.
  QH198.H3.C23 2013
  333.95′15309969—dc23
                                        2012044810

University of Hawaiʻi Press books are printed on acid-free paper and meet the guidelines for permanence and durability of the Council on Library Resources.

Designed by Julie Matsuo-Chun
Printed by Thomson-Shore, Inc.

To my children, **Paul** and **Ellen**,
and their generation.

I suspect there are two categories of judgment which cannot be delegated to experts, which every man must judge for himself, and on which the intuitive conclusion of the non-expert is perhaps as likely to be correct as that of the professional. One of these is what is right. The other is what is beautiful.

—**Aldo Leopold**

CONTENTS

xi	*Acknowledgments*
xvii	*Introduction: Restoring a Rainbow*
1	PART 1. IF YOU PLANT IT, WILL THEY COME?
3	1. Journey to Hakalau
19	2. Place of Many Perches and Hooves
37	3. Science to the Rescue?
50	4. Laulima
67	5. Place of Many New Perches and Fewer Hooves
81	PART 2. RESTORATION ROUNDUP
83	6. Kill and Restore: Hawai'i Volcanoes National Park
102	7. The *Pū'olē'olē* Blows: Dry Forest Restoration at Auwahi, Maui
130	8. Turning Hands: Limahuli Botanical Garden, Kaua'i
165	PART 3: HERDING CATS WITH LEAF BLOWERS
167	9. Multiple Perspectives
187	10. Nature Is Dead. Long Live Nature!
215	*Bibliography*
227	*Index*
	Color plates follow page 144.

ACKNOWLEDGMENTS

Mahalo to the many people in Hawai'i who inspired me, challenged my ideas, answered my questions, accommodated my endless needs, and helped me understand and tell the tales presented in this book. These people include Adam Asquith, Donna Ball, Randy Bartlett, Tom Bell, Dave Bender, Wayne Borth, Marie Bruegmann, David Burney, Katie Cassel, Mick Castillo, Chuck Chimera, Melissa Chimera, Colleen Cole, Susan Cordell, Ellen Coulombe, Julie Denslow, Saara DeWalt, Leilani Durand, Mary Evanson, Kerri Fay, Tim Flynn, Charlotte Forbes, Betsy Gagne, Bill Garnett, Christian Giardina, Jim Glynn, Don Goo, David Godale, Lisa Hadway, Richard Hanna, Audrey Haraguchi, Eileen Harrington, Roger Harris, Pat Hart, Stephen Hight, Baron Horiuchi, Flint Hughes, Jim Jacobi, Jack Jeffrey, Tracy Johnson, Jordan Jokiel, Boone Kauffman, Kapua Kawelo, Creighton Litton, Rhonda Loh, Lloyd Loope, David Lorence, Art Medeiros, Tami Melton, Theresa Menard, Trae Menard, Nancy Merrill, Alex Michailidis, Matthew Notch, Becky Ostertag, Steve Perlman, Lyman Perry, Karen Poiana, Linda Pratt, Peter Raven, Don Reeser,

Joby Rohrer, Darren Sandquist, Paul Scowcroft, Phyllis Somers, Hannah Springer, Bryon Stevens, Bill Stormont, Jarrod Thaxton, Mike Tomich, Pat Tummons, Tim Tunison, Alan Urakami, Erica von Allmen, Warren Wagner, Rick Warshauer, Dick Wass, Steve Weller, Chipper Wichman, Haleakahauoli Wichman, Kawika Winter, and Ken Wood.

Thanks to the people and institutions that provided much-appreciated support and guidance during the earliest stages of this project. The 2004 Bread Loaf Writers' Conference in general and my participation in William Kittredge's nonfiction workshop in particular (superbly assisted by Kristin Henderson, Sebastian Matthews, and my fellow students) were especially influential and helpful. Plattsburgh State University's Institute for Ethics in Public Life provided a stimulating and productive environment in which to write and think throughout my time there as a fellow. I would like to thank Barbara Dean at Island Press for generously allowing me to adapt some material from my earlier work, *Intelligent Tinkering*, for this new book which draws on my experiences in Hawai'i with a different focus. As the present book developed, Heather Fitzgerald, Ibit Getchell, Eric Pallant, Cherry Racusin, Tom Vandewater, Paddy Woodworth, and the students in several classes at Brevard College read one or more chapters and offered encouragement and constructive criticism.

Dave Bender, Emory Griffin-Noyes, Jack Jeffrey, Fernando Juan, Rhonda Loh, Sierra McDaniel, Don Reeser, Phyllis Somers, Forest Star, Kim Star, Erica von Allmen, Mark Wasser, and Kawika Winter kindly donated their own photos or helped me locate others to illustrate the restoration stories profiled in this book. I thank the many staff members and volunteers at Auwahi, Hawai'i Volcanoes National Park, Hakalau Forest National Wildlife Refuge, and Limahuli Garden, who provided critically important firsthand information, observations, and unpublished documents. I am especially grateful to Arthur Medeiros, Nancy Merrill, Don Reeser, Pat Tummons, and Erica von Allmen for their foundational interviews and/or writings. Jack Jeffrey, Lloyd Loope, Art Medeiros, Erica von Allmen, and Kawika Winter read portions of the manuscript and provided invaluable corrections, comments, and updates. Of course, any remaining errors in this book are mine alone.

I am deeply indebted to Andre Clewell and Randall Mitchell for reading the entire draft of an earlier version of this manuscript and providing substantial expert criticism and support. Nadine Little, my editor at the University of Hawai'i Press, provided steady and insightful guidance during the book's latter stages.

Finally, for their only slightly flagging patience and support throughout the process of my writing another book, my family deserves the greatest thanks of all, so Anne, Ellen, Paul, Rhea, and Seymour: *Mahalo nui loa!*

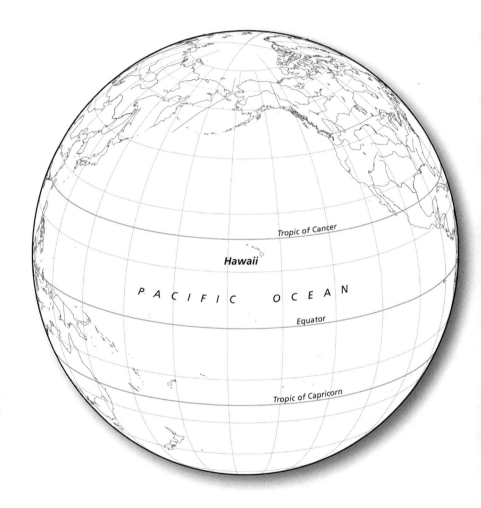

Geographic isolation of the Hawaiian Islands.

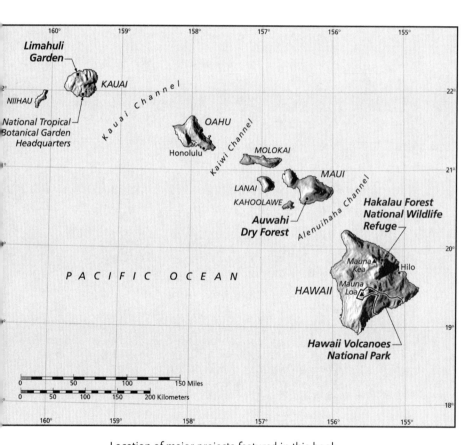

Location of major projects featured in this book.

INTRODUCTION

Restoring a Rainbow

"Whenever I give a conservation-oriented talk," a senior colleague once told me, "regardless of where I am, when I ask who in the audience has had firsthand experience with extinction, virtually everyone who raises their hand winds up talking about Hawai'i." What happened to the po'ouli is a particularly dramatic example of one such Hawaiian extinction story. This stocky, secretive, sparrow-sized honeycreeper was discovered in 1973 by three University of Hawai'i students in a remote high-elevation rain forest on East Maui. The po'ouli's appearance and behavior were so unique that it was eventually placed into its own taxonomic category. Among other things, it was the only Hawaiian bird known to heavily forage on native tree snails, yet another formerly diverse and abundant group of species that is now mostly on the edge of extinction or already extinct. At the time of its discovery, scientists estimated there were fewer than two hundred po'ouli left; a few years later it was formally listed under the newly created US Endangered Species Act.

The *po'ouli*'s numbers subsequently declined due to a laundry list of factors that are all too familiar to those who work in Hawai'i: habitat loss and degradation; noxious alien species; introduced predators; exotic diseases. By the mid-1990s, only one male and two females were left, and the *po'ouli* was dubbed the "world's rarest bird." In a last desperate attempt to save it, scientists tried to capture and artificially breed those last three birds. Unfortunately, they managed to catch only the male, which they brought to the Maui Bird Conservation Center in September 2004. This old bird was missing an eye and had recently contracted avian malaria, an exotic disease spread by nonnative mosquitoes.

After a series of increasingly aggressive interventions by veterinarians failed, what was probably the last of the *po'ouli* (the two females had not been spotted in nearly a year) died in captivity on Friday, December 26, 2004. "I don't think it was a mistake," Eric VanderWerf, the US Fish and Wildlife Service's Hawaiian bird recovery coordinator, said at the time. "If we had left the birds where they were, the species would certainly [have] gone extinct. It may anyway, but I think that was really the only option we had at the time." VanderWerf appears to have been right on both counts: captive breeding probably was their only viable option, but because no *po'ouli* have ever been seen or heard again, it almost certainly went extinct anyway.

I first landed in Hawai'i in 1996 on a postdoctoral fellowship at the National Tropical Botanical Garden on Kaua'i. Throughout this fellowship, and then later as a research ecologist for the US Forest Service based on the island of Hawai'i, I was immersed in the scientific and conservation communities' efforts to better understand, preserve, and restore Hawai'i's native species and ecosystems.

At first, I thought Hawai'i was a hopeless ecological disaster. In the beginning, it often felt as if Hawaiian conservation was largely a matter of documenting the steady deterioration of the native fauna and flora and bracing for the next *po'ouli*-like disaster. But I gradually came to see that there was much more to this story. For instance, despite all the extinctions, there are still about twelve thousand extant Hawaiian species that are unique to this archipelago, and more new and intriguing species are discovered every year. Perhaps even more importantly, I eventually discovered that numerous individuals and organizations were quietly implementing successful and inspiring on-the-ground conservation projects across the islands.

For example, the Plant Extinction Prevention Program focuses on rescuing Hawaiian species with fewer than fifty remaining individuals (so-called PEP species). In 2010 alone, this program protected 116 PEP species, 101 of which were federally endangered. Protection efforts include managing threats from nonnative species such as feral pigs and rats, propagating and replanting PEP plants in the wild, and surveying new areas for additional populations of PEP species. "We are the arms and legs of the Endangered Species Act," Joan Yoshioka, PEP's coordinator, told me. "Our staff routinely prevents extinctions and even rediscovers and saves species that were presumed already extinct. Their dedication and often heroic efforts in the field also build trust, goodwill, and critically important partnerships with landowners, local communities, and a wide variety of other organizations."

I have chosen to devote the majority of this book to a handful of these kinds of more hopeful stories for three reasons. First, the depressing ones are already relatively well known; the press seems to love the "death of the last *poʻouli*" stories far more than stories such as, "PEP saves another species." Second, because these success stories demonstrate that at least some of Hawaiʻi's remaining native biodiversity can be preserved and restored, I hope they will help inspire us to do more before it really does become too late. Finally, I believe these stories are actually far more compelling and important than the sad ones and that they can teach us even more about ourselves and our complex and changing relationship with nature.

After telling a few of these tales, I delve more explicitly into the intellectual and philosophical issues associated with designing and implementing real-world conservation projects. I am well aware that some members of the Hawaiian conservation community in particular and the larger environmental movement in general tend to be impatient with and even hostile toward the academics, whom they perceive as whiling away their time pondering esoteric topics such as the nature of nature. Indeed, after spending more time in the Hawaiian conservation trenches, I went through a period in which I felt like shaking some of these people by their smug collars and yelling, "Wake up! While you have been enjoying your little intellectual adventures, we just lost another irreplaceable native forest that provided crucial habitat to several critically endangered species. How 'bout grabbing a rifle and a shovel and doing some real work for a change?"

However, as more time passed, I found myself becoming less sure about exactly what we could and should be doing to save Hawai'i. Given our often pathetically limited knowledge and resources, which species and ecosystems should we focus on? What should we do to them, and who gets to decide? What role should science play in on-the-ground conservation? Why do so many seemingly like-minded people within the conservation and scientific communities often disagree with each other so passionately? How do we get more of the general public to care about ecology and conservation?

Trying to resolve these kinds of questions ultimately became at least as important to me as the "real work" of shooting feral pigs and saving endangered native species. The evolution of my thinking was well encapsulated by a contemporary environmental historian who noted, "At a time when threats to the environment have never been greater, it may be tempting to believe that people need to be mounting the barricades rather than asking abstract questions about the human place in nature. Yet without confronting such questions, it will be hard to know which barricades to mount, and harder still to persuade large numbers of people to mount them with us. To protect the nature that is all around us, we must think long and hard about the nature we carry inside our heads."

Thus in addition to grappling with the more concrete and physical world of resource management, modern conservationists must also navigate their way through the seemingly more esoteric labyrinth of topics such as humanity's relationship to nature. In other words, not only do we have to figure out how to, say, shoot animals a and b, poison plants c and d, and reintroduce species e and f back into the wild, but we must also develop and defend a coherent intellectual rationale for why we want to do these often costly and controversial things in the first place.

Hawai'i's many paradoxes provide a fascinating microcosm in which to examine both the practical and intellectual dimensions of applied conservation. Despite representing a mere 0.2 percent of the

US land area, this archipelago contains all of the Earth's climates, most of its ecosystems, and some of its most culturally and racially diverse human communities. Despite their relatively low overall biological diversity, three-quarters of all of America's bird and plant extinctions have occurred within these islands. Despite the fact that portions of this paradise have become high-end cosmopolitan playgrounds for the ultrarich and famous, much of Hawai'i remains poor, insular, and at best unreceptive to the relatively affluent and often foreign environmentalists trying to save its unique species and ecosystems. Finally, despite Hawai'i's many severe ecological and socioeconomic challenges, if we Americans truly believe in preserving our remaining biological diversity, we cannot afford to practice ecological triage and give up on our Fiftieth State. This is because all four of its counties now rank in the top five US counties for federally endangered plants and animals, and forty-eight of the fifty-nine endangered species listed by the Obama administration in the past two years have been Hawaiian plants and birds. In fact, these islands now contain more endangered species per square mile than anywhere else in the world.

The Hawaiian Islands consist of a long chain of mostly extinct volcanoes that stretches about 1,500 miles from the island of Hawai'i to Kure Atoll. Beyond Kure, the chain continues in a northwesterly direction for perhaps another 2,500 miles as a series of submerged volcanoes, or seamounts. Each Hawaiian island was created by a stationary "hot spot" that pushes magma up through the Pacific Plate. This tectonic plate drifts in a northwesterly direction at an annual rate of about 3.5 inches and has slowly rafted away each of the new volcanoes that formed over this hot spot. Thus as one moves up the chain in a northwesterly direction, each volcano (or "island" while it lies above the ocean and "seamount" when it sinks below) is progressively older and more weathered. The oldest known seamount, near the Aleutian Islands, is about 80 million years old, while the actively erupting Lō'ihi Seamount, eighteen miles off the island of Hawai'i's southeast coast, is predicted to emerge as the newest Hawaiian island within the next 200,000 years.

The dynamic history of this archipelago may be divided into three periods: prehuman (before the first humans reached these islands fifteen hundred to eight hundred years ago), prehistoric (from the time the first people landed in Hawai'i until westerners arrived in 1778), and modern (1778 to the present). The ecology of these islands during the vast prehuman period was quite different than what followed in the prehistoric and modern eras: islands rose out of the sea; living creatures colonized them and evolved into strange and beautiful species; islands eroded and sank back into the sea. While many of the details of these processes remain shrouded in mystery, we do know that at one point several species of giant flightless birds dominated these islands and catalyzed a cascade of ecological and evolutionary interactions that almost certainly never occurred before and will never happen again.

It is tempting to view this prehuman period as a kind of golden age in which a rainbow of delicate species and pristine ecosystems coexisted harmoniously in a perfect and unchanging Garden of Eden. However, modern ecologists tend to view the prehuman world of Hawai'i and everywhere else as being neither harmonious nor unchanging. On the contrary, since the mid-twentieth century they have increasingly viewed nature as a complex, ever-evolving entity whose species and ecosystems change in unpredictable and often drastic ways whether or not humans are around. For instance, natural "disasters" such as volcanic eruptions, hurricanes, and tsunamis must have caused enormous physical and biological changes throughout Hawai'i's prehuman period. New "alien" species occasionally arrived, displaced the existing "native" species, and directly and indirectly caused them to go extinct. Over time, rain forests morphed into dry forests and then back into rain forests again as local and regional climates changed. Thus what the primeval, "real" Hawai'i looked like depends on when and where we look.

Hawai'i's prehistoric period began about a thousand years before Columbus landed in North America, when Polynesian navigators reached these islands in their double-hulled sailing canoes. We still don't know exactly why they sailed from their ancestral islands across thousands of miles of open ocean without modern navigational instruments and ultimately discovered the most isolated islands in the world. Was it escape from overcrowding, oppression, or war at home? Or simply the sheer adventure of it?

After such a long and difficult journey, these spectacularly beautiful and benign islands must have seemed like paradise. Ironically, however, these sailors would have quickly discovered that this paradise contained few if any plants worth eating: due to their extreme geographic isolation, none of Polynesia's familiar and staple food plants were present, and none of the native species provided much edible vegetable carbohydrate. But they did find substantial sources of marine animal protein, such as mollusks, fishes, and sea turtles, and unimaginably abundant flocks of birds. Some of these birds were enormous, and because there had been nothing around to eat them throughout the prehuman period, over evolutionary time they lost the ability to fly and the need to be wary of predators. In other words, they were quite literally giant sitting ducks.

Over the next several hundred years, these first human settlers made numerous back-and-forth journeys between Hawai'i and their ancestral Polynesian islands. On the return voyages they brought back their most important animals, such as dogs, pigs, and chickens (some argue that rather than being an unintentional stowaway, the Polynesian rat was also deliberately brought over as a source of protein for the dogs, sport hunting, or simply as a reminder of home, as apparently was the case for several nonutilitarian tree species). They also brought many of the food plants we think of today as quintessentially "Hawaiian," such as banana, breadfruit, and coconut. But some time after AD 1200, these two-way voyages mysteriously ceased, and Hawai'i once again became isolated from the rest of the world.

Some have suggested that the end of these ancient back-and-forth journeys was at least partially caused by the overharvesting of the massive native koa trees the Hawaiians needed to construct their large voyaging canoes. Proponents of this theory point out that when Captain Cook reached the island of Hawai'i in 1779, his officers spotted thousands of koa canoes in Kealakekua Bay alone. It is unclear how many of the giant koa "canoe trees" still remained in Hawai'i's forests at that time, but their complete absence became painfully obvious near the end of the twentieth century when the Polynesian Voyaging Society decided to reenact those original sailing journeys. Although this group strived to use as many traditional materials as possible to construct their famous *Hawai'iloa* canoe, after an extensive search failed to find a single koa tree that was sufficiently healthy and large enough for their canoe's hulls, they ultimately used two Sitka spruce trees from Alaska.

Others believe the cessation of those round-trip sailing voyages was more likely due to the massive reduction of birds in Hawai'i and throughout Polynesia that occurred during that period. Millions of seabirds once lived in the Pacific, and many of these birds regularly migrated from archipelago to archipelago across vast stretches of open ocean. Archeological and fossil evidence strongly suggests the Hawaiians eventually decimated these birds by eating them in large quantities and altering much of their coastal habitat. Even if they had strictly enforced a *kapu* (taboo) on killing these birds once they noticed their increasing scarcity, predation from their introduced pigs, dogs, and rats most likely would have continued to devastate them. Some scholars have concluded that the ancient sailors depended on those migrating flocks for crucial navigational assistance, and it is therefore no coincidence that their round-trip voyages stopped around the same time that they wiped out their seabirds.

Whether or not this hypothesis is correct, it highlights the tragic fate of birds in the Pacific in general and within the Hawaiian Islands in particular. In addition to the drastic reduction of their primordial abundance, paleontologists estimate that more than two thousand species of land birds (about 20 percent of all the bird species described for the entire planet!) have become extinct since humans colonized the Pacific Islands. When the Polynesians first reached Hawai'i, there were at least 140 species of birds within these islands. The majority of these, including all 80 land bird species, were found nowhere else in the world. In the one thousand or so years since the Polynesians arrived, 71 of the islands' 113 unique bird species have gone extinct, and 33 of the surviving 42 species are endangered. Ten of these 42 species have not been seen in over forty years; like the *po'ouli*, they are almost certainly already extinct.

The Polynesians ultimately created a remarkably sophisticated and thriving civilization in Hawai'i. Living literally in the middle of nowhere, with no alphabet, Arabic number system, wheels, metals, clays for pottery, or draft animals, they were by necessity intensely practical and observant people. Beyond these kinds of basic facts, however, much of Hawai'i's prehistoric human and natural ecology remains unclear.

Hawai'i's modern period abruptly began in 1778 when Captain James Cook became the first white man to reach the islands when he accidentally discovered them while searching for a northwest passage between England and the Orient. Cook and his successors deliberately and unintentionally brought many new species to the islands that turned out to be notorious ecological wrecking machines, including cattle, goats, rodents, mosquitoes, ants, and a world-class collection of noxious weeds.

Captain James King, who sailed on Cook's voyage to Hawai'i in 1778–1779, estimated that there were 400,000 people living within the entire archipelago at that time. Credible contemporary estimates of Hawai'i's total prehistoric population range from 150,000 to more than 1.5 million people. According to the US Census, the total population of the State of Hawai'i in 2010 was 1,360,301. Thus it is conceivable that there were about as many people living within these islands before the arrival of the first westerners as there are today. However, the decline of the indigenous population after Cook's arrival is tragically clear: the 1831–1832 census reported 130,000 total Native Hawaiians; by 1876 this population fell to an all-time low of only 54,000. Some scholars also believe that Hawai'i's prehistoric population had peaked well before the eighteenth century and was thus already in decline before Cook's arrival due to widespread environmental degradation. They support this argument in part by citing archaeological evidence that suggests human settlements in some of the islands' ecologically marginal areas were abandoned before the eighteenth century.

Until relatively recently, most modern, "civilized" people have grossly underestimated the sophistication of indigenous cultures and their ability to profoundly alter their environments. For example, many of Hawai'i's first Western explorers were struck by the treeless and barren character of the islands' coastal lowlands. Cook and his officers commented on the "woods that so remarkably surround this island [the island of Hawai'i] at a uniform distance of four to five miles from the shore." Twenty years later, Captain George Vancouver similarly noted that on Kaua'i, "the sides of the hills extending from these [taro] plantations to the commencement of the forest, a space comprehending at least one half of the island, appeared to produce nothing but a coarse spiry grass from an argillaceous soil, which had the appearance of having undergone the action of fire."

Some of these westerners concocted elaborate theories to explain how phenomena such as this absence of coastal trees had naturally occurred in Hawai'i. Yet we now know that throughout much of the prehuman period, most of these islands were densely forested all the way down to their shores. These early Europeans were apparently unable to imagine that throughout the prehistoric period (and continuing right under their noses well after Western contact), the Hawaiians, like indigenous peoples throughout the tropics, had cleared and burned their lowland forests and converted them into cultivated grasslands, agricultural plantations, and thickly settled villages. In fact, we now know that some of the world's most biologically diverse and "pristine" tropical rain forests are literally growing out of the ruins of once extensive and sophisticated pre-European civilizations.

Thus for most of the modern era, if they thought about it at all, westerners have assumed that the ecological impact of the Native Hawaiians was trivial. However, a wealth of evidence gathered over the past several decades has increasingly challenged this "noble savage" view of prehistoric cultures in Hawai'i and throughout the rest of the world. Indeed, some scholars now believe that humans have caused waves of environmental destruction and species extinctions virtually every time we settled in a new place for the last fifty thousand years. Quantifying the overall environmental impact of prehistoric people is a difficult and often contentious task, but at least for relatively well studied species such as Hawaiian birds, most scientists are now convinced that the majority of these extinctions occurred during the prehistoric period. (Although the ancient Polynesians must have also consumed large amounts of ocean animal protein, their effects on the native marine fauna appear to have been quite modest, probably because of this ecosystem's greater size and resiliency.)

Not surprisingly, some have angrily denounced and contested these kinds of conclusions. Like the descendants of other indigenous peoples, some contemporary Native Hawaiians view such "accusations" as yet another wave of haole (foreign) oppression and racism and a self-serving attempt to justify the past and ongoing ecological sins of Western civilizations. Perhaps due to a combination of guilt over their ancestor's horrendous treatment of native peoples and their lands, a desire to support today's indigenous peoples, and an overly romantic vision of the prehistoric past, some nonindigenous people also vehemently reject these claims.

Nevertheless, there now appears to be a growing acceptance among indigenous cultures that their ancestors did significantly alter their environments and drive other species into extinction, although their interpretations of this history often differ substantially from those of nonindigenous people. For instance, when I asked a prominent leader within the Native Hawaiian community about these prehistoric bird extinctions, I received the following response:

> It's important to remember that the lens through which this history is being viewed is an external lens—the same people that brought us all those devastating alien species and caused and are causing so much environmental and cultural destruction are now saying, "You fools! Look what you did, and what you're doing today!" So if there's a defensiveness among my people, that's one of the places where it's coming from. But if that defensiveness expands to say that we lived in harmony always, that's ignorance. If we as Native Hawaiians think like that, we're not being introspective; we're not being confident enough of our place in the landscape.
>
> I believe the *kapu* system was put into place by my ancestors in part because they saw diminishing returns from the land around them. We can look at some of the Hawaiian creation epics and find descriptions that to me are quite in keeping with people who are keen observers of their world and the causes and effects of some of the events that are occurring on the landscape. I can only imagine what the food and fuel and building material needs were for the folks of old! But if we don't recognize our impact as people of Hawaiian descent, we're not learning the lesson of colonialism that we're so busy getting on our soapboxes and decrying. Whoever we are, if we cannot be introspective and make corrections of ourselves, we're just doomed to embitterment and finger-wagging.

I know that the Indians on the US mainland had long been tinkering with the landscapes that the Europeans thought were "expansive wildernesses." They deliberately created environments that were suitable for the species they desired. I have no doubt that we were doing the same thing here in Hawai'i, and on a first-name basis! These various plants and birds and fish were the body forms of our ancestors, who were elevated to the deity status. So, all those bird extinctions . . . yeah, we did it! But in the same way that we can't judge our ancestor's activities with twenty-first-century eyes, neither can we in the twenty-first century necessarily assume their values as our own today. That lens-looking works both ways.

Modern environmentalists must often consciously choose the lens through which they view the natural world. For example, in Hawai'i we could decide to consider the prehistoric Polynesians as being entirely separate from and outside of nature. Viewed through this lens, these people and everything they brought with them were the first alien species to reach these islands. Because their arrival and subsequent ecological impacts were therefore unnatural, if our goal is to preserve and restore these islands to their natural condition, we should strive to erase as much of their environmental footprint as possible.

On the other hand, we could choose to think of those first humans as being as much a part of nature as the birds and turtles and spiders that colonized these islands before them. Viewed through this lens, the species these people brought with them in their canoes and all the ways they subsequently modified the islands to suit their needs were entirely natural phenomena. Consequently, attempting to erase their footprints would be as misguided as trying to remove a native bird and all of the myriad ways in which it has altered the islands' ecology (the plants it originally brought to Hawai'i as seeds embedded in its muddy webbed feet, the disturbances created by its nest making and seed dispersal activities, and so on).

Through what lens should we view Captain Cook and all of the people who followed him to Hawai'i in the modern age? If we see them as being just another part of nature, then all of the species and materials and technologies they brought to the islands are, too, and we should think of Honolulu's skyscrapers and traffic jams as being just as natural as the patches of remnant "native" forests that still haunt portions of this island today. If we choose to view these modern people as being entirely unnatural, does that mean we necessarily have to view the prehistoric Hawaiians as separate from nature as well, or can we legitimately classify different groups of people differently?

Or would it be wiser to employ a more nuanced and piecemeal approach? We could decide that, say, based on their present ecological effects, some of the species brought to the islands by prehistoric and modern peoples are natural and others are unnatural aliens that should be removed. However, if we rigidly employed this model, we would at least occasionally find ourselves in the awkward position of eradicating some beloved and long-established Polynesian species while preserving other species that recently arrived from ecologically and culturally distant places such as Europe and South America.

Because different people and organizations view the world through radically different lenses, they often wind up with radically different ideas for how we should interact with and manage the natural world today. Yet at least in Hawai'i, one thing we can all agree on is that these islands' prehuman and prehistoric ecosystems have been fundamentally and irreversibly altered by both "natural" and human forces. We will never see those vast flocks of flightless birds and swarms of brightly colored tree snails again.

What can and should we do about this? Humanity's perception of and reaction to its transformation of nature has varied considerably across different cultures, races, social classes, places, and times. For example, throughout New England's precolonial period, most Europeans were proud of the ways in which their culture "tamed" what they perceived as an unruly and evil landscape. As early as 1653, one historian approvingly viewed the transformation of a "remote, rocky, barren, busy, wild-woody wilderness" into a "second England for fertilness" as a product of divine guidance and inspiration. As more time passed, however, some of the descendants of those early Europeans began to see the ecological changes they had wrought more as a fall from the Garden than the planting of one. Writing in

Massachusetts in 1855, Henry David Thoreau passionately expressed this sentiment:

> When I consider that the nobler animals have been exterminated here,—the cougar, panther, lynx, wolverine, wolf, bear, moose, deer, the beaver, the turkey, etc., etc.,— I cannot but feel as if I lived in a tamed, and, as it were, emasculated country.... I take infinite pains to know all the phenomena of the spring, for instance, thinking that I have here the entire poem, and then, to my chagrin, I hear that it is but an imperfect copy that I possess and have read, that my ancestors have torn out many of the first leaves and grandest passages, and mutilated it in many places. I should not like to think that some demigod had come before me and picked out some of the best of the stars. I wish to know an entire heaven and an entire earth.

Today we know that the ecological situation in Thoreau's New England was far more complex than he realized. Many of the components of the entire poem that Thoreau lamented had been torn out by his ancestors had been created and maintained by the extensive activities of the Native Americans who preceded them, and we can only guess at how those indigenous peoples may have tamed and emasculated the "pristine wilderness" they first encountered.

Since our emergence as a populous species, we have apparently always directly and indirectly affected the ecology of at least our local environments. However, it does not follow that vast ecological destruction is an inevitable by-product of all human cultures, or that one kind of ecological change is necessarily no better or worse than another. Some people found ways to live on this Earth for far longer than others without significantly degrading their own ecosystems or subjugating someone else's. Given the increasingly global and detrimental environmental effects of our modern civilizations, a major challenge for today's conservationists is to develop more sustainable yet feasible ways of interacting with nature.

American environmentalists once focused most of their conservation efforts on creating more wilderness, a term poetically defined by the US Wilderness Act of 1964 as "an area where the earth and community of life are untrammeled by man, where man himself is a visitor who does not remain." Like it or not, that era is gone—we can no longer literally or figuratively put a fence around an area and walk away for three reasons. First, there are no untrammeled areas left to preserve (and we seem determined these days to trammel our existing natural areas as fast as possible). Second, even our wildest remaining species and ecosystems are now directly affected by humans because our footprints are no longer confined to our cities and farms—the average polar bear must now contend with human-created problems such as DDT and PCBs, hunting, alien species, and of course all those melting icebergs. Finally, some scholars now argue that the very idea of wilderness is a misguided abstraction based on a Garden of Eden-type myth. They point out that the 9 million acres of land that were instantaneously "preserved" by the Wilderness Act had long been (and in some cases were still being) intentionally managed by indigenous people. To paraphrase one contemporary Native American writer, wilderness is what you call a place when you don't know its stories.

Nevertheless, despite all the intellectual, physical, ecological, and socioeconomic challenges associated with modern conservation, some people have concluded that doing something is surely better than doing nothing. As different as these individuals and their institutions may be, they are all more or less united by their desire to consciously and proactively manage the Earth, both to preserve its biological diversity and to create more meaningful and sustainable relationships between people and nature. In other words, they are committed to the emerging paradigm of ecological restoration.

An extraordinarily broad array of people is now designing and implementing ecological restoration programs all over the world. This group includes natural and social scientists, philosophers, indigenous peoples, landscape architects, natural resource managers, engineers, farmers, and community activists. A similarly broad array of institutions is also actively involved in ecological restoration. For example, one of my main research and restoration sites in Hawai'i was managed by a working group comprised of representatives from more than twenty-five organizations that included government agencies, botanic gardens, private industries, nonprofit environmental groups, and academic institutions. Like so many other restoration programs, we also depended on the critical support of volunteers who ranged from local elementary schoolkids and Native Hawaiian teenagers to real estate agents.

Consequently, the conservation success stories I tell in this book revolve around these kinds of complex ecological restoration programs. In part 1, I tell the in-depth story of the Hakalau Forest National Wildlife Refuge's ongoing efforts to restore thousands of acres of degraded pastures on the island of Hawai'i back to the diverse native rain forests that once dominated this area and sheltered a suite of native birds now on the brink of extinction. Along the way, I provide an overview of Hawaiian natural and cultural history, biogeography, and evolutionary biology.

To offer a flavor of the diversity of restoration projects and approaches across the state, in part 2 I tell three shorter stories from three different islands. "Kill and Restore" highlights the US Park Service's efforts to control alien species and reestablish native species and ecosystems within their vast Hawai'i Volcanoes National Park on the island of Hawai'i. "The *Pū'olē'olē* Blows" follows one charismatic scientist's efforts to restore an ecologically and culturally important forest on the island of Maui that most experts believed was unrestorable. "Turning Hands" tells the story of the biocultural restoration of a thousand-year-old taro plantation at the Limahuli branch of the National Tropical Botanical Garden on the island of Kaua'i.

To more explicitly investigate the diverse and often conflicting motivations, philosophies, and on-the-ground strategies of the people involved in these kinds of ecological restoration programs, part 3 begins with excerpts from my interviews of a broad swath of the larger Hawaiian environmental community. Finally, I use these interviews and the previously presented restoration stories as springboards for a more general discussion of the human/nature relationship, the power and limitations of science, and the future of conservation within and beyond the Hawaiian Islands.

Part 1   **If You Plant It,
         Will They Come?**

1   JOURNEY TO HAKALAU

I dashed through the pounding rain into the Ford Bronco, turned the wipers and defrosters on high, and headed out for another day of research at the Hakalau Forest National Wildlife Refuge. Although this refuge is only about fourteen miles northwest of Hilo, I knew it would take me over two hours to get there due to the rain, fog, and primitive roads. When I added in another forty-five minutes of driving within the refuge to get into and out of my field site and the hour and a half round-trip journey from my home in Volcano Village to the office, I realized I was in for another six-hour-plus day behind the wheel. Ironically, despite my love of nature and the outdoors and aversion to cars and bureaucracies, since moving to the island of Hawai'i (the "Big Island") to work as a research ecologist for the US Forest Service in 1997 I seemed to spend an ever-increasing portion of my life driving around one of the most isolated islands in the world in a gas-guzzling government vehicle.

I turned away from the ocean and began the long slow climb to Hakalau on the Saddle Road, what was then a narrow two-lane highway that runs from Hilo Bay across the center of the island between the state's two largest volcanoes (13,677-foot Mauna Loa to the south and 13,796-foot Mauna Kea to the north). Named for the broad saddle that lies between these two massive mountains, this highway was hastily constructed by the US Army in 1942 for military purposes. Until recent improvements, it was a winding, poorly paved road with blind curves that weren't properly banked, and the central and eastern portions are often shrouded in rain and heavy fog.

After the tiny "Last Chance" convenience store a few miles up, there isn't another shop, gas station, or stop sign along the Saddle's remaining fifty miles of pavement. Yet because the rental car companies forbade most of their customers to drive on it and because civilization has not penetrated this far inland, the vast and often surreal landscape on both sides of this highway offers some of Hawai'i's least populated and most beautiful vistas.

I failed to resist the twin temptations of speeding and looking at the scenery as I snaked my way through the dripping forests and black lava fields. As usual, my feeble excuses were (1) I needed to get about a day and a half's work done at Hakalau; and (2) there was so much to see! In stark contrast to the vast majority of drivable places in the rest of the state, most of the vegetation along the twenty-mile stretch of rain forest that extends from upper Hilo to the barren lava flows farther up on the Saddle is native.

Unlike continental tropical rain forests, which may have several hundred different species of canopy trees, Hawai'i's rain forests are dominated by just two: 'ōhi'a lehua (*Metrosideros polymorpha*) and koa (*Acacia koa*). This dramatic difference in diversity is largely due to the extreme difficulty of dispersing to and establishing in these islands. For instance, the roughly one thousand species of native flowering plants found in Hawai'i today evolved from about 275 original immigrants. Assuming this occurred over a period of approximately 30 million years (the age of Kure Atoll, the oldest extant island within the Hawaiian chain), this works out to only one successful plant colonization every hundred thousand years. The native bird fauna, which evolved from about twenty original immigrants, apparently achieved less than one successful colonization every million years. Thus in these and many other groups, speciation within the Hawaiian

Islands far exceeds the rate at which new species immigrated to this archipelago from elsewhere.

Before humans, of course, the only way to reach these islands was by wind, wing, or water. ʻŌhiʻa, by far the most common native tree in Hawaiʻi, produces enormous numbers of tiny, lightweight, aerodynamic seeds that are rapidly dispersed by the wind. The ancestors of this species probably originated within the volcanic habitats of New Zealand and then hopscotched their way to Hawaiʻi by using other Pacific islands as stepping-stones. Surprisingly, however, less than 2 percent of Hawaiʻi's flowering plants appear to have dispersed here by air; seeds carried by birds account for nearly 75 percent of the successful colonizations, and another 23 percent probably got to Hawaiʻi by direct flotation or rafting on floating debris.

As a result of this extreme geographic isolation, Hawaiʻi's terrestrial native biota is both depauperate (species poor) and disharmonic (weird). Many of the familiar plant and animal groups that are important components of continental ecosystems were virtually or completely absent throughout Hawaiʻi's evolutionary history simply because they either could not cross that vast, empty ocean or could not establish a reproducing population on the islands if they did manage to get here. Thus the archipelago does not have any native representatives from such normally ubiquitous groups as gymnosperms, mangroves, mammals (aside from bats), reptiles, amphibians, or ants. And even though the Big Island is sometimes called the "Orchid Island," there are only three inconspicuous native orchids in the state's entire flora. In this case, since orchids are among the most diverse of all tropical plants and produce tiny, dustlike seeds that are widely dispersed by the wind, it is probably safe to assume their seeds must have repeatedly landed in Hawaiʻi. Yet even if these seeds had managed to germinate and grow to maturity, the absence of their specialized insect pollinators would have made sexual reproduction impossible for the great majority of orchid species. This same scenario also probably explains why Hawaiʻi has no native fig or banyan species.

The few species that did manage to disperse to and establish within the Hawaiian Islands were an odd mix of colonists from all over the globe. These "waif elements" suddenly found themselves in a world without most of the species that had shaped their ecological niches and evolutionary trajectories back home. The interaction of the islands' climate and topography also created sometimes wildly different habitats within short distances of one another. As a whole, these factors ultimately led to the evolution of about seventy land birds, five thousand insects, a thousand flowering plants, and eight hundred species of snails found nowhere else in the world. While these numbers pale in comparison to the species richness typically found in mainland tropical ecosystems, the evolutionary shifts and species radiations undergone by some of these original colonists are arguably the most spectacular in the world.

For example, the handful of birds that reached Hawai'i from elsewhere eventually occupied and adapted to habitats ranging from well above the tree line on the highest volcanoes to the remote atolls hundreds of miles northwest of the major high islands. Presumably because the costs of flying outweighed its benefits in a middle-of-nowhere-world without ground-dwelling predators, at least ten species of flightless ducks and geese and over a dozen species of flightless rails descended from those first avian colonists. More than fifty-five species of Hawaiian honeycreepers apparently evolved from a single rosefinch species that colonized the islands between six and seven million years ago. The strikingly different beaks of this guild of birds—short, thick bills for crushing hard seeds; thin, straight bills for extracting insects from twigs and bark; long, thin, decurved bills for probing flowers for nectar—suggest that, much like the patterns Darwin famously encountered among the Galapagos finches, this Hawaiian species radiation was driven by adaptations for different diets and foraging techniques.

Although less charismatic than the native avifauna, Hawai'i's pomace or vinegar fly fauna may be the world's most spectacular example of adaptive radiation in the animal kingdom. More than 350 species of flies in the genus *Drosophila* appear to have descended from one or two ancestors that reached Hawai'i around 10 million years ago. About a hundred species of these picture-winged flies have dark spots on their wings and unique structures and courtship behaviors. While it is unclear how many drosophilids went extinct without ever being discovered, the small populations and limited distributions of many of the extant species suggest there may once have been considerably more Hawaiian species in this genus.

The most famous example of adaptive radiation in the Hawaiian flora is undoubtedly the magnificent silverswords. Genetic studies have shown that all thirty-six members of this group probably evolved from a single founding population of California tarweeds that reached Hawai'i around 6 million years ago. The descendants of this ancestral tarweed eventually dispersed to and thrived in habitats ranging from the wettest place on earth (Mount Wai'ale'ale on Kaua'i) to the arid leeward slopes of the Big Island and Maui. To adapt to these radically different niches, these species diversified into compact cushion plants, rosette shrubs, perennial vines, shrubs, and large woody trees. Yet despite these extreme morphological, ecological, and

Mauna Loa or Ka'ū silversword (*Argyroxiphium kauense*). *National Park Service*

physiological differences, all of these species are still so similar genetically that it is possible to pollinate any member with any other and produce viable seed.

Because the Hawaiian biota contains so many world-class examples of diversification and adaptive radiation, it is tempting to imagine that every species that reached these islands subsequently underwent dramatic ecological and evolutionary changes. Indeed, many popular and technical articles about "Hawaiian-style" evolution understandably focus on the most sensational stories, such as the fact that a single plant immigrant that arrived 13 million years ago appears to have subsequently radiated into 126 separate species (one-eighth of the entire native flowering plant flora!). But these examples are the exception, not the rule. For instance, in the case of the flowering plants, more than two-thirds of all the plant immigrants apparently established in these islands with either zero or only one speciation event. Why a select few colonists exhibited such explosive diversification and speciation while the majority underwent only modest changes remains an important yet unanswered question in evolutionary biology.

As I wound my way up the Saddle, the rain and mist began to thin and I could just make out the lichen-covered crowns of the koa trees. Koa is considered the monarch of the Hawaiian forest; there are still koa that are over one hundred feet tall and wide enough to drive a car through. Next to ʻōhiʻa, this species is the most common native tree in Hawaiʻi, occurring on all the main islands except Niʻihau and Kahoʻolawe. In prehistoric times, koa thrived under both dry and moist conditions from near sea level to eight thousand feet, but today they survive mainly in upland areas, and only about 10 percent of all the original koa forests are left. Yet despite the centuries of massive deforestation and ecological abuse, at least in a few places such as the Hakalau Forest National Wildlife Refuge, there is good reason to believe that we might be able to reverse the tide and see these magnificent koa forests rise again.

Koa is a good example of a species that successfully colonized and spread across the archipelago without exhibiting any spectacular ecological or evolutionary changes. Although individual trees can look quite different from each other, there are only two native *Acacia* species in Hawaiʻi, and while both of these species are unique to these islands, they are still closely related to another *Acacia* species found in the Mascarene Islands.

Even with my eyes (mostly) glued to the wet and winding road, I could still discern three distinct layers of native vegetation thriving beneath the koa. Scraggly, wizened ʻōhiʻa trees dominated the midstory. These trees were much smaller and less elegant than the stately koa towering above them, but their dense clumps of crimson, pompom-like flowers popping out of the mist made it hard to look at anything else (see Plate 1). I also couldn't help gawking at how variable different individuals or even different parts of the same individual could be. As Sherwin Carlquist put it in his classic *Hawaii: A Natural History,* "There is no such thing as a typical *ohia.*"

ʻŌhiʻa is a prime example of a plant that adapted to Hawaiʻi's physical and ecological diversity by evolving extensive intraspecific variability. Although the taxonomy of *Metrosideros* is still problematic,

Layers within a Native Hawaiian rain forest. *Jack Jeffrey*

it is generally accepted that there are only five native species and eight varieties in Hawai'i. *M. polymorpha* is by far the most common; its specific epithet—*polymorpha* or "many forms"—is, if anything, an understatement. This single species still occurs from near sea level to over eight thousand feet in niches ranging from baking hot lava flows to cold and spongy high elevation bogs, where it typically exhibits a bonsai-like form in which a mature "tree" may be only six inches tall. In other habitats, 'ōhi'a may appear as an epiphyte, scrubby shrub, or very large canopy tree.

Most 'ōhi'a flowers are red, but they may also be various shades of orange, yellow, or white, and their leaves can be smooth, hairy, round, cuplike, or trembling like a quaking aspen's. In transition areas such as the border between a bog and a mature rain forest, individuals from the two adjacent habitats may hybridize and produce offspring with intermediate characteristics. Shortly after I first arrived in Hawai'i, I learned that whenever I encountered a weird-looking native plant whose identity baffled me, my best bet was to confidently declare it to be just another 'ōhi'a.

Below the midstory 'ōhi'a lay another picturesque layer of shrubby trees, such as 'ōlapa (*Cheirodendron trigynum*, a member of the ginseng family), famous for their slender green leaflets that do a little hula dance (*lapalapa*) in the breeze, and one of my favorites, the giant *hāpu'u* tree ferns (*Cibotium* spp.). These species can grow over thirty feet tall and send out twelve-foot wands of feathery, Dr. Seuss–like fronds that look like they would have been the perfect snack for herbivorous dinosaurs. Finally, in the gaps between the *hāpu'u* lay a smothering groundcover of dripping *uluhe* (*Dicranopteris linearis*) ferns. The first time I encountered an impenetrable wall of these ferns while bushwhacking through a remnant native forest, I was sure this species was just another noxious weed.

As I drove through the rain and mist, the highway began to weave in and out of some more sparsely vegetated lava flows. The barest of these were dominated by lichen-encrusted rocks and waves of dead or dying 'ōhi'a. When large numbers of canopy 'ōhi'a trees began dying within this island's montane rain forests in the 1960s, scientists assumed that an epidemic disease was responsible for their widespread mortality. However, intensive research during the 1970s and 1980s suggested that this dieback seemed to be a recurring natural phenomenon in aging 'ōhi'a forests. While various ecological and climatic factors may accelerate this process, they now believe this stand-level 'ōhi'a dieback is primarily caused by the aging of individual, even-aged cohorts of trees.

Because of their copious production of readily dispersed seeds, 'ōhi'a is often the first native plant to colonize open areas such as new lava flows. Like many other early successional pioneering species, 'ōhi'a are shade-intolerant and thus do not establish within closed-canopy forests. Consequently, mature 'ōhi'a stands often consist of trees that germinate, grow, mature, and die together. In the evolutionary past, as the canopy died back, the increasing sunlight reaching the forest floor would have triggered the establishment of a new cohort of 'ōhi'a and other native plant species. Today, however, this cycle can be broken by shade-tolerant alien species that invade healthy, closed-canopy native forests and shade-intolerant alien species that out-compete 'ōhi'a and other native species in the high-light environments created by canopy diebacks and other disturbances such as lava flows, fires, and hurricanes.

Looking out the window at all those dead trees, I wondered whether I was witnessing the very last act of an ecological drama that had been playing in these islands for eons. Would all subsequent generations of plants in this region be dominated by alien species? Would my memories of these low-elevation native forests eventually become yet another nostalgic tale of the good old days in Hawai'i?

As I drove out of the last forested pockets and into the wide open lava-field plains, the sky abruptly cleared and both volcanoes burst into view (see Plate 2). To my left I spotted the Mauna Loa Observatory glinting in the sun. At eleven thousand feet in the middle of the Pacific Ocean, this station has been collecting and analyzing samples of some of the purest air on the planet since 1957 as part of NOAA's Climate Monitoring and Diagnostics Laboratory. To my right I glimpsed a few of the baker's dozen telescopes sprinkled around the summit of Mauna Kea. Because the conditions on top of this volcano are ideal for stargazing, many of the world's best telescopes are there.

The considerable disturbance created by building and operating

Hāpu'u tree fern (*Cibotium* spp.) colonizing barren lava flow. *Jack Jeffrey*

these astronomical observatories has angered some Native Hawaiians who consider the summit region of Mauna Kea (White Mountain) the sacred home of Pol'iahu, the snow goddess. Many environmentalists also fear this development will hurt its aeolian alpine lava ecosystem. This unique ecological community is sparsely vegetated by lichens, algae, and mosses that shelter an improbable native fauna that includes a wolf spider, centipede, flightless moth, and the famous *wēkiu* bug (*Nysius wekiuicola*). The genus *Nysius* radiated into at least twenty-six species in Hawai'i, and all of these except the *wēkiu* and its close relative on Mauna Loa make their living by feeding on the seeds of native plants. But the rice grain–sized *wēkiu* lives exclusively within the few feet of loose volcanic cinders that lie on top of Mauna Kea's permanent subterranean ice sheets, and it preys on dead birds and insects that are blown up the mountain from below.

I turned off the Saddle near its 6,578-foot crest, then turned again a few miles later onto a dirt road and put the Bronco in four-wheel drive. Although I was now only about a dozen miles from the Hakalau Forest Refuge, it would take another forty-five minutes to get there due to the rutted, washed-out road and occasional need to brake for crossing cattle.

As I bounced along, I saw that the surrounding pasture was full of mullein (*Verbascum thapsus*), a biennial Eurasian weed that sends up spikes of yellow flowers from a thick basal rosette. There was some debate within the Hawaiian conservation community over whether and to what extent it was spreading and causing trouble. Some botanists thought we should aggressively go after it now, before it was too late, while others felt it was *manini* (small change).

I had a secret fondness for mullein; it was one of the only plants I recognized when I first landed in Hawai'i because it was common in my native New England, and I'd often used its large velvety leaves as toilet paper on camping trips. Moreover, while I was working as a visiting professor at Kenyon College in Ohio prior to moving to Hawai'i, one of my least favorite students had confessed that all he got out of smoking the mullein I had shown his ecology class was a sore throat and massive headache. For some reason, he thought I had claimed that Native Americans regularly dried and smoked this species to clear out their lungs and get high.

In a few minutes I drove out of the grass and mullein pastures and into a vast field of gorse (*Ulex europaeus*). Native to Western Europe, this spiny, leguminous shrub was inadvertently introduced to Hawai'i by the wool industry around the turn of the twentieth century. Apparently it was kept more or less in check by the sheep until the ranchers switched to cattle after wool prices plummeted during the Great Depression. When mature, this species becomes a wicked tangle of sharp needles that are virtually impenetrable to humans and animals alike. Gorse now dominates more than twenty thousand acres of montane pastures on the eastern slopes of Mauna Kea in what is easily the nastiest weed infestation I have ever seen (see Plate 3). Anyone who thinks Mother Nature is an all-benevolent goddess who should always be allowed to take her course should spend a few minutes within a dense, twelve-foot tall thicket of gorse.

Gorse (*Ulex europaeus*). Jack Jeffrey

Early one morning as I was driving down to Hilo, I noticed a glow off to the north. Turning to look more closely, I saw that a large section of Mauna Kea's eastern flank was illuminated by a series of bright, criss-crossing orange lines. Could this volcano be erupting after 4,500 years of dormancy? Baffled, I pulled off the highway and studied the lines more closely until I finally realized that what was glowing was not molten lava but burning gorse.

The Mauna Kea gorse infestation adversely impacts three major ranch operations that have been leasing pastures for generations from the Department of Hawaiian Home Lands, a state agency responsible for managing lands in the interest of the Native Hawaiian community. When their leases expire, these ranches are legally required to return the land to the state in a condition equivalent to how it had been when they first began leasing it, which in this case means gorse-free. In an effort both to comply with this legal requirement and to reclaim thousands of acres that have become worse than useless for their cattle, the ranchers attacked gorse from the air with herbicide-spraying helicopters and from the ground with bulldozers and controlled burns. But their war has not gone so well, and some of the key players have grown increasingly frustrated with the bureaucratic restrictions surrounding the use of fire as a gorse-control tool. At least a few of these people seemed to have a professional and personal desire to see this shrub burn in hell, and it was widely suspected that one or more of these individuals was behind the gorse "wildfire" I had seen on my drive down to Hilo.

Gorse is, in fact, extremely flammable—it has been burned for hundreds of years in Europe—and may in the short term be effectively controlled by fire. However, this species, like many other noxious invasive weeds, produces enormous numbers of long-lived seeds that can rapidly germinate and establish following fires. Burning gorse on Mauna Kea also means destroying any co-occurring native species, as well as running the very real risk of fire spreading into adjacent areas such as the ecologically priceless remaining old-growth native forests within the Hakalau Forest Refuge.

In the 1970s, temporary control of this gorse infestation had been achieved through the large-scale application of 2, 4, 5-T (the active ingredient in Agent Orange). However, after the Environmental Protection Agency banned this herbicide, the gorse population quickly began to recover. In an attempt to manage and coordinate an effective control program, the Mauna Kea Soil and Water Conservation District created a Gorse Task Force that included affected land owners, ranchers, and representatives from relevant state and federal agencies. This task force explored a variety of gorse management strategies, including biological, chemical, and mechanical control programs, intensive sheep and goat grazing, controlled burns, and reforestation.

The US Forest Service was brought in to help develop and implement the biological control and reforestation components of this program. Biocontrol, as it is commonly abbreviated, is the art and science of using one or more deliberately introduced alien species to control the distribution and abundance of another problematic alien species. The theory behind this fight-fire-with-fire approach is that species are often limited in their home countries by other coevolved native species such as predators, herbivores, and pathogens. Biocontrol practitioners thus attempt to find these kinds of species from the target alien's place of origin and import them to the nonnative countries where it is causing problems. In this case, the Forest Service ultimately released a slew of biocontrol agents against the gorse, including a web-building spider mite, seed-eating weevil, soft-shoot moth, and rust-inducing fungus. Their early reforestation trials also suggested that both koa and an alien pine species might be able to effectively overtop and eventually shade out or at least suppress the underlying gorse.

When I attended one of these task force meetings shortly after I started working for the Forest Service, I was struck by how its tone seemed to rival the nastiness of the gorse infestation itself. For example, one researcher bitterly complained that the controlled burns and bulldozers were wiping out his biocontrol agents. He was convinced that if everyone else would just back off and butt out, his critters would eventually control the gorse on their own. This provoked an equally irate response from another member, who claimed that the hundreds of thousands of dollars spent on biocontrol had been a total waste. If he had been given a fraction of that money for his no-nonsense mechanical and chemical control programs, he said, his crews would have "eradicated the bastard long ago!" Another participant smugly remarked that when faced with too many lemons, a wise man will make lemonade. He then argued that we should build a power plant that could generate electricity by continually burning truckloads of sustainably harvested gorse.

As the meeting wore on, I detected an inverse relationship between the extent of peoples' on-the-ground experience battling invasive species and their confidence in their own ability to conceptualize the problems and devise the most effective solutions. I later discovered that some academics and bureaucrats with little to no real-world experience had similarly dogmatic ideas about how agencies such as the Forest Service should conduct their basic ecological research and applied natural resource management programs. Yet much like the people doing most of the talking at the gorse meetings, these individuals tended to have little interest in or respect for the experiences and insights of the people who spent their lives in the trenches.

Most of the fieldwork associated with the Forest Service's gorse biocontrol and reforestation projects was performed by two technicians named Donovan Goo and Alan Urakami. Both of these men were born and raised in Hawai'i, both are part Native Hawaiian, and, after serving in Vietnam, both returned to the Big Island and began working for what became the Forest Service's Hilo-based Control of Non-Indigenous Plant Species Team. When I first met them during my job interview, I was immediately struck by their unique combination of warmth, gentleness, humility, and regular outbursts of uproarious and contagious laughter. Sadly, I later discovered that their extensive knowledge and insightful ideas were unappreciated or outright ignored by many of the scientists and bureaucrats they worked for and interacted with. But it was not until several years later, after I had gained a measure of their trust and some painful on-the-ground experience myself, that I began to grasp just how brutal their gorse biocontrol and reforestation fieldwork must have been.

One day as we were driving through the gorse on our way to Hakalau, Alan looked out the window and announced that he had devised a plan for what to do with all this land. "Look," he said, "the state has way more prisoners than it can handle, so it has to ship them off to places like Texas because no one in Hawai'i wants a new prison in their backyard. But sending them to the mainland costs a lot of money, and it creates a real hardship for the friends and families of the prisoners. So why not build a new prison out here, where nobody would care?"

"Well," I said, "I thought this was supposed to be Hawaiian Home Lands. Isn't the state eventually supposed to turn this over to guys like you and Don who have been waiting for your land for decades?"

Alan slapped his thigh and chuckled while Don roared.

"Oh yeah," Alan said. "Why you think I worked so hard all those years killing gorse and planting koa? I think my deed is supposed to come at the end of the month."

"Really?" Don deadpanned, putting his hands on his jaw in mock shock. "Too bad for you—they said they put mine in the mail yesterday!"

"No, seriously," Alan said finally, his voice hoarse from laughing. "They could build a prison out here—that land will never be good for nothing again—and have the prisoners slowly chip away at the gorse and plant koa. Since they got nothing else to do with their time anyway, they could do what Don and I have been saying for years: wherever a weed has made it through, remove and replace it with a native."

"Yeah," Don replied, wide-eyed, "and they wouldn't have to build no fences around the prison either. Anybody tries to make a break for it and survives all that . . . they *deserve* to be free!"

Despite decades of meetings, research, on-the-ground containment efforts, and optimistic papers and presentations, more than forty thousand acres of land on the Big Island and Maui remain infested with gorse, and smaller populations have been reported on other islands. Some worry that gorse, like so many of Hawai'i's other noxious alien species, has already fallen into the category of "yesterday's problem." The war against such alien species has in fact often followed a familiar pattern: after an initially galvanizing "call-to action" period, Hawai'i's overwhelmed and beleaguered alien species research, management, and outreach communities are prematurely pushed or pulled into confronting some newer and even more pressing alien invasion.

2 PLACE OF MANY PERCHES AND HOOVES

Shortly after finally emerging from the worst of the gorse infestation, I turned onto another rough road that took me to the entrance gate of the Hakalau Forest Refuge. I shut off the engine and jumped out, eager to inhale fresh mountain air after the long, bone-rattling drive in our moldy truck. The outside world was perfectly still, and the dew on the tall pasture grasses sparkled in the morning sunlight. The thick clouds over Hilo were beginning to break, and behind them were reassuring blue skies over the ocean for as far as I could see. I turned toward the rising sun, closed my eyes, took several deep breaths, and slowly rotated my head and neck around my stiff and anxious shoulders.

I drove into the refuge, then got out and carefully relocked the gate—the last thing Hakalau needed was more ungulate ingress. As I rolled past the modest administration buildings, I stopped to watch a gaggle of banded *nēnē* (*Branta sandvicensis*) foraging in the cut grass just off the road. This handsome goose is the official state bird of Hawai'i and closely resembles its presumed ancestor, the extant North American Canada goose (see Plate 4). When Captain Cook arrived in Hawai'i, there were about twenty-five thousand *nēnē* left in the islands, but by 1951 fewer than thirty remained. Subsequent intensive captive breeding and release programs have at least temporarily rescued Hawai'i's last surviving goose from extinction, and today there are over two thousand free-living birds in the wild. However, without intensive human assistance within their managed nesting areas, these "wild" populations inevitably decline due to their degraded habitats, vulnerability to exotic predators, and reduced genetic diversity.

The Hakalau Forest Refuge was created in 1985 with the purchase of two parcels of former ranchland totaling 8,313 acres. Although much of this acreage contained some of the finest remaining stands of native montane rain forest, this purchase was literally "for the birds." In the mid-1970s, the US Fish and Wildlife Service (FWS) realized that little was known about Hawai'i's forest birds: nobody could say with any confidence where the birds were, how many were left, or what was happening to their populations and habitats. To address these questions, the FWS initiated the Hawai'i Forest Bird Survey and hired a group of dedicated young people to rigorously survey the entire archipelago. Due to the rugged, remote nature of its higher elevation forests, it took this team over three years just to survey the Big Island. When the entire Bird Survey was finally completed, upland portions of eastern Mauna Kea emerged as a clear hotspot for endangered native birds. The subsequent creation of the Hakalau (Hawaiian for "place of many perches") Forest National Wildlife Refuge under the authority of the Endangered Species Act was the first time in the United States that a refuge was established to preserve native forest birds and their habitat. (Most of the other refuges were created to provide habitat for game animals such as migratory waterfowl.)

Today this refuge comprises almost thirty-three thousand acres between 2,500 and 6,600 feet. Because the lands below 4,000 feet receive over 250 inches of annual rain, these areas mostly consist of bogs, matted ferns, and scrubby *ōhi'a* forests. Above 4,500 feet, rainfall declines to about 150 inches per year, and in some places giant old-growth koa and *ōhi'a* stands anchor a diverse, closed-canopy rain forest. Jack Jeffrey, Hakalau's biologist, once told me about a three-meter-diameter koa on the refuge that required nine hand-holding kids to encircle, as well as a 1.5-meter-diameter 'ōhi'a tree. Unlike the speedy koa, which can grow two to five centimeters in diameter in a single year on the refuge, Jack said that on average their 'ōhi'a grow only about one to two millimeters per year. "Does that mean," I asked him, "that that 1.5-meter 'ōhi'a tree might be 1,500 years old?" Jack threw up his hands. "Who knows?"

Although this mid-elevation, more mesic region provides the best habitat for the native endangered birds, it is also the most degraded vegetation type at the refuge, having been decimated by over two hundred years of cattle grazing, logging, fires, invasive weeds, and feral pigs. "When the cattle started to breed here," Jack explained, "it was like a giant salad bar. Nothing prevented them from eating all the native plants, and after that regeneration was kept in check by grazing animals." Thus today the best mesic forests at Hakalau are primarily found in the areas where cattle grazing was restricted by rugged topography, a protective underlying substrate of jagged 'a'ā lava, or both. Above six thousand feet, the annual precipitation further decreases to less than one hundred inches and the rain forest peters out into open fields of dense alien pasture grasses.

The Hakalau Forest National Wildlife Refuge can be seen as a microcosm of the history and future of Hawaiian biodiversity as a whole; depending on one's perspective, there is legitimate cause for both optimism and despair. On the bright side, so many unique, beautiful, and fascinating species still exist here, and there is no reason to believe we cannot save the great majority of them. The refuge now provides critical habitat for fourteen species of native birds, twenty-nine rare plants, and the state's only native mammal, the endangered Hawaiian hoary bat.

Yet on the darker side, so much has already been lost, and much of what remains is hanging by a thread. For example, seven species that were part of Hakalau's avifauna when Captain Cook landed on the Big Island are now extinct, and eight of the birds still with us are federally endangered. One of these species, a finch-billed, yellow-headed honeycreeper called the 'ō'ū (*Psittirostra psittacea*), has not been seen on the island since 1986. Several of the plant species are down to their last few individuals, and nearly half of the rare plant flora is endangered. One native mint species was presumed extinct until it was rediscovered on the refuge in 1990. Little is known about the bat other than it appears to be an elusive, nocturnal, and insectivorous species.

Despite their diversity and complexity, most large-scale restoration projects can be broken down into three major, often intertwined management goals: (1) Remove or mitigate the source or sources responsible for the past and present ecological degradation; (2) facilitate the recovery and establishment of the most important target species; and (3) carefully monitor what happens and refine the management activities accordingly.

At Hakalau, as is the case throughout the Hawaiian Islands, the primary biological sources of ecological degradation were and are ungulates and weeds. Because ungulates often facilitate the establishment of weeds by dispersing their seeds while simultaneously destroying the native vegetation, the first step toward ecological restoration in Hawai'i is almost always ungulate control, and the first step toward ungulate control is almost always erecting a fence, because, as Jack Jeffrey put it, "Without fencing, there is no restoration."

Despite their explicit federal mandate to preserve and restore native birds and their habitat, however, the refuge's efforts to fence out their resident ungulate populations have been hindered in two major ways. First, Hakalau is surrounded by state and privately owned lands teeming with ungulates, weeds, and other alien species. The state's Piha Game Management Area actually bisects the refuge and, as its name implies, it is managed to *maintain* large populations of feral pigs. Thus no matter how successful the refuge's conservation and restoration programs may be, abundant populations of some of Hawai'i's most destructive alien species may always be literally waiting at their gates. Second and perhaps even more importantly, hunters and their allies have consistently opposed the federal government's efforts to eradicate the pigs *within* the refuge.

Most scholars agree that the ecological impact of the ancient Polynesians was largely confined to the lowlands. Except for occasional forays into the vast primeval upland forests for special purposes such as harvesting giant koa trees for voyaging canoes and catching birds for their colorful feathers, the ancient Hawaiians mostly left

Effects of cattle grazing on a native rain forest. *Jack Jeffrey*

Rain forest recovery following ungulate exclusion. *Jack Jeffrey*

these sacred areas alone. However, in 1793 and 1794, Captain George Vancouver brought cattle to King Kamehameha and made him promise to ban the killing of these animals for at least ten years. By the early 1800s, there were reports of cattle destroying private gardens and tearing up patches of commercial sugarcane. By the 1820s, "immense herds of wild cattle" were observed on Mauna Kea; in 1829, a single hunter reportedly killed forty thousand cattle on the Big Island alone.

By the 1830s, large numbers of wild cattle roamed all the major Hawaiian Islands. Ranchers had also created their own domesticated populations out of these wild stocks that they grazed on the royalty's lands in exchange for money or a share of their rapidly expanding herds. Yet while the sale of beef, hides, and tallow became major economic commodities, the agricultural and ecological damage inflicted by cattle also became increasingly problematic. For example, on the Big Island, the town of Waimea was once described as a "gardened landscape" that included "evergreen hills and extended plain diversified with thick wood, open pasture, low shrubbery and fruitful plantation." But in 1847, a Waimea resident noted that two-thirds of the town had been converted into government pastureland, and "people are compelled to leave their cultivated spots and seek distant corners of the woods beyond the reach of the roaming cattle, sheep, and goats . . . but the cattle follow, and soon destroy the fruit of their labors."

The "distant corners of the woods" rapidly became sparser and more remote. In 1856, the editor of the *Sandwich Islands' Monthly Magazine* wrote that cattle had destroyed Waimea's "thick wood," and "at this moment they swarm in the thick jungle that covers windward or eastern slope toward Hamakua. They are now gradually destroying this, and thousands of old dead trees both standing and lying prostrate form the present boundary of these woods and exhibit the mode in which the destruction is effected; for whilst the old trees die of age, no young ones are seen taking their places, as during the last 30 or 40 years, the cattle have eaten or trodden them down." In Waimea and across the archipelago, older residents complained that the widespread destruction of vegetation by cattle had caused their local climates to become windier, colder, and drier. Once dependable streams and irrigation networks began to run dry, and water shortages on Oʻahu became common.

By the 1860s, the Hawaiian sugar industry had become a dominant economic and political force. The plantation owners realized that unless something was done, the ongoing destruction of the upland forests would also destroy the source of the water and wood they needed to grow and process their sugarcane. They began to aggressively lobby for legislation to promote forest preservation and restoration, but most of their early efforts were thwarted by the ranchers, who controlled much of the remaining forests and increasingly depended on them to provide the water and rich fodder that was no longer available on their dry and overgrazed pastures. The sugar plantations responded by fencing tens of thousands of acres above their fields, driving out the cattle, and planting hundreds of thousands of trees within these fenced reserves. In the twentieth century, Hawai'i's territorial government finally began to protect the islands' forests and established a network of public forest reserves that eventually included more than 1.2 million acres.

In 2002, Patricia Tummons, editor of the state's leading environmental newsletter, *Environment Hawai'i,* wrote an article entitled, "Hawaiian Forests: How Do You Celebrate a Century of Loss?" Her essay began with the following reflection:

> A century ago, tens of thousands of acres of Hawaiian forests stood dead and dying, but stood nonetheless. The snags created what one observer called a ghost forest, and riding through it must have been as haunting as a midnight walk through a cemetery over the graves of one's ancestors. We have lost the ghost forests. Replacing them is stubble called pasture that is scarcely able to support the cattle and other livestock, which, along with their owners, were largely responsible for the forests' destruction. As the trees, both living and dead, have disappeared, we have lost more than just the forests. We have been robbed of a measure of our loss that, four generations ago, moved men to heroic action. We no longer have the markers and headstones to show us the sheer, vast numbers of the fallen. And in their passing, we have been left more removed than ever from the reproachful incitement to outrage and to action that spurred Hawai'i's

first foresters to undertake swift, bold measures to repair the islands' scarred uplands.

Tummons then states that the heroic actions that established Hawai'i's forest reserves were motivated by the urgent need to protect the water supplies for the agricultural industry and the residents of Honolulu. Because the modern concepts of valuing native biodiversity and preserving endangered species did not exist at that time, those early foresters understandably restored Hawai'i's denuded lands with fast-growing exotic trees such as eucalyptus. Rather than condemn these men for establishing forests of alien species across the archipelago, she feels we should be grateful to them for successfully retaining and restoring some of Hawai'i's most important watersheds. But Tummons then points out that continuing to focus on the utilitarian value of the forests creates a risk of overlooking the urgent need to preserve the uniqueness of the islands' native forests. Like many people within Hawai'i's conservation community, she argues that the state should be managing its lands to preserve and restore native ecosystems and rare and endangered species, rather than sacrificing them to the numerically and economically weak yet politically strong ranching and hunting lobbies.

Hawai'i's scientific and environmental communities are comprised of a diverse and often cantankerous group of organizations and people. I have even observed a few "debates" within these communities degenerate into nasty shouting matches and personalized attacks. But when the discussion turns to what is the single greatest threat to Hawai'i's remaining native rain forests, virtually every scientist and environmentalist gives the same one-word answer: pigs.

The ecological impacts of pigs include both painfully obvious and more subtle effects. Let loose in the rain forests, they behave like, well, pigs—turning over the earth in search of worms and slugs, wallowing in the mud, and knocking over and trampling everything in their way. Due to the absence of predators and competitors and their prolific reproductive rates, feral pig populations can double within a single year and exceed seventy-five individuals per square kilometer of Hawaiian rain forest. One study calculated that the pigs within the Kīlauea State Forest Preserve on the Big Island turn over half of the diggable area in less than a year. I have seen large patches of native forests so thoroughly churned up by pigs that they looked as if someone had camped out for weeks with a rototiller and gone berserk. Pigs can also catalyze a cascade of other less obvious yet equally devastating effects. For example, they often transport noxious alien plant seeds or spores on their hairy coats or within their feces deep into native forests and then facilitate the establishment of these weeds by destroying the native vegetation and disturbing the soil. Their rooting and defecation can also change soil nitrogen availability and nutrient cycling processes in ways that favor the establishment of alien plant and invertebrate species. In addition, pigs love to knock over native tree ferns and devour their starchy interior cores. The resulting hollowed-out trunks may then collect water and provide key breeding sites for the alien southern house mosquito (*Culex quinquefasciatus*), which is a vector for the alien protozoa that cause avian malaria and the virus that causes avian pox. These two diseases in turn have decimated the native avifauna and continue to severely limit their distribution and abundance across the archipelago.

Not surprisingly, many environmentalists dream of ridding Hawai'i of every last feral pig (and cow, deer, sheep, and goat while they're at it). What makes this vision particularly tantalizing is that unlike so many of the other destructive alien pests such as rats, weeds, insects, and microorganisms, this dream is at least theoretically possible. With sufficient money, labor, and political will, we probably could shoot, snare, trap, or poison every last hoofed mammal in the state. But not everyone would like to see these animals eradicated; in fact, the "numerically weak yet politically strong" hunting lobby (there are roughly ten thousand hunters in Hawai'i) has successfully lobbied the state to manage much of its land explicitly *for* these ungulates.

This environmentalist versus hunter battle has proven to be among the most polarizing and intractable land management issues in the islands. Like many conservation conflicts, this fight is exacerbated by profound cultural differences between the two opposing sides. At the extreme environmental end of the continuum lies the stereotypical wealthy, white outsider who values biodiversity and wilderness. At the other extreme end is the less affluent, nonwhite local who, like his father and grandfather before him, both enjoys hunting

Feral pig damage. Note knocked-down and hollowed-out tree fern. *Jack Jeffrey*

Alien mosquito on native 'apapane (*Himatione sanguinea*). *Jack Jeffrey*

and values a freezer full of free meat. Many Native Hawaiians also love to hunt and strongly identify with the importance of swine in traditional Polynesian culture.

While no one argues over the Western origin of all of Hawai'i's other ungulate species, the ancestry and historical ecology of today's feral pigs remain unclear and contentious. We know the Hawaiians' ancestors transported their Polynesian pigs to these islands in their sailing canoes, and it is clear that these swine were extremely abundant by the time of Western contact. For instance, Cook's ships in 1778 and 1779 were furnished with six hundred pigs, and early westerners reported that well over a thousand of them might be cooked and eaten at the consecration of an important temple. Yet it is also well established that these prehuman pigs weighed only about a third as much as the 150-pound feral pigs that ravage Hawai'i today. What is less clear is why those Polynesian pigs were so small. Was it due to their low-protein diets (before Western contact, there were few sources of protein in native ecosystems), the way they were farmed, their underlying genetics, or a combination of these and other factors?

An even more important and controversial topic is the extent to which these Polynesian pigs escaped domestication and formed abundant and ecologically destructive feral populations. Some researchers believe that the Hawaiians carefully managed and controlled their pigs, and thus any damage caused by these animals would have been largely confined to the lowland areas adjacent to their settlements and agricultural fields. However, others have argued that these pigs were much more free ranging and widespread, and that they may have caused substantial damage to native species ranging from plants to birds to sea turtle eggs.

Nevertheless, virtually all of today's environmentalists and scientists view feral pigs as just another invasive Western species that has no business being in Hawai'i. Some also point out that the costs associated with hunting pigs and other ungulates (four-wheel drive vehicles, hunting dogs, guns, and so on) can greatly exceed the value of the resulting "free meat." For their part, many hunters and some Native Hawaiians argue that pigs and other feral ungulates do not destroy native ecosystems and believe the scientists who study these animals are biased and untrustworthy. Some even claim that ungulates help Hawaiʻi's ecosystems by keeping down the weeds and sowing the seeds of native plants.

In one of the more bizarre political alliances ever formed, Hawaiʻi's hunting lobby has often garnered the invaluable support of animal rights groups such as the People for the Ethical Treatment of Animals. These groups oppose what they perceive as inhumane ungulate control measures such as snaring and helicopter shootings. Hunters in turn object to these methods because they can effectively control and even eradicate ungulate populations and because they result in a lot of "wasted" meat.

This fight between the environmentalist and the hunting/animal rights communities has smoldered for generations, and it has occasionally erupted into violence. Conservation-oriented ungulate management efforts have frequently been blocked politically and undermined in the field by people vandalizing fences and boycotting efforts to hunt or drive out the animals. Consequently, vast sections of the islands are still either explicitly managed for "sustained yield" ungulate hunting or simply not managed at all. Inevitably, these lands turn into weedy areas dominated by alien species that can coexist and even thrive with dense ungulate populations.

In theory, it would seem that the two sides could agree on a commonsense compromise: manage the degraded areas for game animals and hunting and reduce or eradicate the ungulates within the best remaining native ecosystems. Yet with some notable exceptions, this has proven exceedingly difficult to do in practice, particularly in places such as the Big Island that have relatively large numbers of people who have grown up hunting more or less wherever they pleased. And to the extreme chagrin of conservationists, hunters often prefer to hunt in relatively intact native ecosystems because they tend to be much more pleasant than the weedy communities that have replaced them.

Shortly after the Hakalau Forest National Wildlife Refuge was established, the Cooperative National Park Resources Studies Unit, National Park Service, and the University of Hawaiʻi performed a collaborative study to assess the biological conditions on the refuge and make management recommendations. Not surprisingly, their report concluded that cattle and pigs were the major agents responsible for the past and ongoing ecological degradation of the refuge. They thus recommended that the cattle should be "removed from the refuge as soon as possible in order to allow forest regeneration to begin" and that "eradication [of pigs] is necessary because the high reproductive potential of feral pigs can result in the rapid repopulation of an area, even when just a few pigs remain." Following the recommendations of this study, the refuge set out to implement some of the standard ungulate control techniques that had been successfully employed elsewhere in Hawai'i, such as dividing the region into more manageable fenced subunits and then eradicating the ungulates within these units via hunting with trained dogs, snaring, trapping, baiting, and the construction of one-way gates.

The refuge began by fencing a 550-acre mid-elevation subunit in 1988, employing professional hunters to systematically kill the pigs and cattle within this unit in 1989, and relying thereafter on snares to catch any remaining fugitives. After subsequently killing a few feral cows, pigs, and dogs, the snares stayed unsprung and the unit was declared ungulate free. Next, they fenced another five thousand acres of degraded mesic forest and used riflemen on helicopter platforms and hunters on the ground to kill or drive out the cattle. Between December 1990 and November 1991, ninety-four cattle were killed from the air and another sixty-six were shot from the ground. After eleven more cattle were taken in the following year, they declared this unit cattle free. Although they did not attempt to systematically hunt the pig population, the refuge staff shot and snared forty-five pigs within this unit in 1991 and another sixty-seven pigs in 1992. Since the snares within this unit subsequently rarely caught any more pigs, they also considered this area essentially pig free.

Of course, the hunting community was not pleased with these results. They knew that if the staff continued to employ this subdivide, fence, and eradicate methodology, there eventually would be no more ungulates left to hunt on the refuge. In an attempt to compromise with the increasingly vocal concerns of these hunters, in 1991 the refuge administration adopted a "Sport Hunting Plan" that called for using public pig hunting as an ungulate control strategy within selected areas. This document stated that this would both "prevent further deterioration of native habitat" and "yield public relations benefits because the hunting public currently perceives the USFWS to be 'anti-hunting' and overly protective of the resources it manages."

Controlling ungulates via public hunting might appear to be the kind of win-win deal that everybody loves: environmentalists get free help and political goodwill by working with local hunters, and the hunters get to enjoy their sport in some of the most beautiful areas remaining in the islands. Richard Wass, who was the Hakalau Forest Refuge manager from its inception until he retired in 2008, is a longtime proponent of public hunting as an ungulate control tool. "Fencing is tough," he told me in 2004, "because it's so expensive. In the early years when we were still fencing relatively accessible areas, it only cost about $5–6 a foot, whereas now it can cost over $10 a foot." Thus another benefit of controlling pigs by public hunting is that the refuge does not have to construct these expensive fences (it cost them $421,114 just to fence their 550- and 5,000-acre subunits) or pay for helicopters and professional hunters. "I know public hunting is controversial," Wass conceded, "but politically it has helped the refuge because we can tell the public we've invited the hunters in to help remove the feral ungulates."

The reason public hunting is so "controversial" within Hawai'i's scientific and conservation communities is that it largely doesn't work. Hunters may initially flock to a newly opened area because of its novelty and abundant game animal populations, and thus for a while the resident ungulates may decrease. However, once the novelty wears off and their prey have declined, hunters usually move on to more fertile grounds and come back only after the depleted animal populations have recovered. They also tend to concentrate their efforts in areas that are closest to the trailheads and easiest to traverse (who wants to slog a 150-pound pig across miles of wet, steep, rugged terrain?). Consequently, decreases in feral ungulate populations as a result of public hunting tend to be both short lived and geographically limited.

Indeed, in the six months after Hakalau's Sport Hunting Plan went into effect in January 1992, data from the staff's feral ungulate transects showed that pig activity within the public hunting region of the refuge declined by almost 25 percent. The next census, however, performed in September 1993, found that pig activity in this unit was slightly higher than it had been before the initiation of public hunting. In contrast, the transect data during this same time period within the five thousand-acre fenced subunit (subjected to ongoing staff hunting and snaring) documented further declines in that parcel's pig population.

Despite these data, the long-standing recommendations of its own biologists, and the professional feral ungulate management community's admonishment not to rely on public hunting, in 1995 Robert Smith, the Pacific Islands Ecoregion manager for the USFWS, sent a memorandum to Richard Wass ordering him to "immediately cease all hunting and/or snaring of pig and feral cattle on the refuge by refuge personnel" until further notice. This edict appeared to be in response to two issues raised by the hunting lobby. First, they were upset because the Hakalau staff was allowed to hunt pigs while off duty in areas of the refuge that were not open to the public. Second, Smith knew that the refuge's ongoing preparation of a comprehensive feral ungulate management plan was going to generate extensive public comments. "I do not believe," Smith wrote in his memo to Wass, "that the Service will appear genuine about considering the input of the public if we have an ungulate control program already underway during the development of the plan and public involvement periods."

The Hakalau Forest Refuge staff met three times over the course of 1995 to discuss its feral ungulate management plans with representatives from the hunting community, state and local politicians, and various federal and state agencies. Not surprisingly, many of these meetings ended up being a replay of the same old environmentalist versus hunter/animal rights battle. In the end, the hunters stated that they would be willing to concede the areas of the refuge that had already been fenced if they were guaranteed that the remaining unfenced areas would be managed for sustained yield hunting. However, to the relief of some exasperated refuge staff and members of the conservation community, Wass ultimately rejected these demands and wrote a letter to one of the major hunting organizations stating that "sustained yield hunting is not compatible with the purpose for which the refuge was established."

Shortly thereafter, Hakalau resumed its ungulate control efforts. By the end of 1997 they had installed a total of forty-four miles of fencing (at a cost of $1,243,627) and created eight separate exclosures that spanned the entire upper elevation portion of the refuge. Then, with about a hundred more miles of fencing needed to finish protecting the forest bird habitat, they ran out of funding. Due to major budget cuts across the USFWS, the refuge has not had enough money to build any more fences or purchase more land since 1997. Budget cuts also eventually forced the refuge to cut their field staff down to only three workers. Outside of their eight exclosures, the rest of the refuge today remains, as Jack Jeffrey put it, "pig city."

They did, however, manage to acquire five thousand more acres in 1997 through the purchase of the Kona Refuge. Although this parcel is located over sixty miles away on the other side of the island, within a totally different ecosystem, it is officially considered part of the Hakalau Forest Refuge. This is because at that time the Newt Gingrich–led US Congress had decreed that there would be no new National Wildlife refuges, but since they apparently forgot to explicitly forbid the expansion of existing refuges, the USFWS somehow managed to convince Congress via some creative mapping to consider this new "Kona subunit" as a natural extension of the Hakalau Forest Refuge. Yet in some ways the joke turned out to be on the FWS, because the surrounding landowners ultimately refused to allow the government to cross their lands, so the only way the staff could access this new subunit was via helicopter.

After ungulates, the next largest headache for restorationists at Hakalau and throughout the state is alien plants. Gorse, for example, blanketed nearly five hundred acres of the upper nonforested areas of the refuge. Both the degraded upper elevation pastures and the relatively intact lower elevation forests also contained numerous weeds known to wreak havoc on Native Hawaiian ecosystems. Ironically, as the staff proceeded to remove the cattle, in some instances their absence created two new problems. First, formerly suppressed noxious weeds such as Florida blackberry (*Rubus argutus*) and banana *poka* (*Passiflora mollissima*, an aggressive vine that can scale and smother even massive old-growth trees) began to explode across some areas of the refuge. Second, during drought periods, the staff had to worry about all that tall, rank, ungrazed pasture grass catching fire and potentially destroying the forests they were working so hard to preserve and restore.

Given the enormity of these and other problems and their extremely limited personnel and budget, the refuge staff realized they could tackle only a small fraction of the tasks that needed to be done. This situation leads to one of the most universal problems faced by conservationists in general: Which problems must be addressed immediately, which can be ignored for the time being, and which might eventually take care of themselves? For example, once the ungulates were finally removed, would at least some of the native understory species be able to recolonize all those abandoned pastures and eventually shade out the weeds? Or conversely, without all those grazers, might some of the pasture weeds now be able to successfully invade even the relatively intact native forests? And what about the needs of Hakalau's crown jewels—the native forest birds for whom the refuge was established in the first place?

As is almost always the case in such restoration projects, the Hakalau staff did not have the luxury of waiting to obtain more complete information before acting—the problems were too urgent and the risks of doing nothing too great. So they analyzed their situation, devised the best plan they could, and sprung into action. One of the first things they did after launching their ungulate control program was to wage war on the gorse infestation before it spread any further. After an intensive effort that included years of herbicide applications and prescribed burns, the original population was largely under control by 1993, and today the refuge is mostly gorse free. As time, money, and field logistics allowed, they also went after some of the other noxious weeds using herbicides, hand grubbing, mechanical removal, and fire. Finally, as the fences went up and the ungulates were removed, the staff began the daunting task of restoring all those miles of degraded treeless pastures.

While many of these initial management activities were by necessity performed on a more or less opportunistic, trial-and-error basis, the staff also established an extensive series of permanent transects across the entire refuge. The purpose of these transects was both to collect data on the distribution and abundance of native and alien plant and animal populations and to assess the ecological effects of their various management activities. They also set up cooperative agreements with other agencies and individual scientists to conduct research in the areas where more knowledge and management recommendations were most urgently needed. As one of those cooperating scientists, I hoped to create a research program that would help them restore those abandoned pastures back into something that resembled the diverse and beautiful native forests they replaced.

3   SCIENCE TO THE RESCUE?

I drove past the *nēnē* and slowly began my descent through the refuge's degraded pastures along a winding and rutted dirt road that led to my prospective research site some 1,500 feet below. Except for the occasional forlorn remnant koa or 'ōhi'a tree and clump of native ferns, these pastures were mostly one big amorphous mass of deceptively innocuous-sounding alien grasses such as kikuyu (*Pennisetum clandestinum*), meadow rice grass (*Ehrharta stipoides*), sweet vernal grass (*Anthoxanthum odoratum*), and velvet grass (*Holcus lanatus*). But inevitably, just when I would start to feel that the grand vision of restoring all those miles of grasses back to native rain forest was a noble but hopeless delusion, the road would take me near one of two different kinds of tree corridors running defiantly across the grasslands.

The first type of corridor contained magnificent old-growth koa and 'ōhi'a trees that offered a tantalizing glimpse of the forest that once grew here. When the thick glaciers high up on Mauna Kea began receding at the end of the last Ice Age, the resulting meltwater carved narrow gulches out of the underlying lava substrate as it raced down the volcano toward the sea. Portions of these gulches apparently proved to be steep and deep enough to protect the vegetation within them from the ungulates, logging, fires, and other forces that eventually destroyed the rest of the forest. Consequently, some high-elevation sections of the refuge today serendipitously contain elongated slivers of relatively intact native forests rising out of the surrounding ecological wasteland.

The existence of these forested gulches was the major inspiration behind the refuge's plan to establish a new network of tree corridors. Just as birds on the mainland often fly along the rivers that flow through highly developed or degraded landscapes, biologists discovered that many of Hawai'i's native forest birds were similarly utilizing Hakalau's old-growth tree corridors to traverse the treeless upper-elevation sections of the refuge. Why not capitalize on this phenomenon by planting long columns of fast-growing native koa trees at regular intervals up and down the pastures across the entire refuge?

After a series of switchbacks, I turned sharply toward the ocean and drove straight down the mountain and parallel to one of these refuge-created koa corridors planted several years earlier. Many of these trees, which were just four- to five-month-old, foot-tall seedlings when transplanted, were now well over fifteen feet high, despite the dense kikuyu grass beneath them. In some ways, I found the existence of these koa even more inspiring than the remnant giants lining the old gulches: instead of the seemingly endless cycle of meetings, reports, and presentations about the importance of implementing landscape-level restoration programs, for once somebody was actually *doing* it! Once again I found myself struggling to keep my eyes off the trees.

The road finally veered away from this koa corridor and plunged down a long, steep slope to a flatter area that had recently been declared ungulate free. I parked next to my favorite 'akala (Hawaiian raspberry) that as usual was laden with ripe purple and golden berries that were larger and sweeter than any other berry I have ever tasted. If I ever act on my dream to develop a commercial raspberry cultivar out of wild (and thornless!) Hawaiian stock, I'm going to start with cuttings from that bush. I gobbled down several pints of berries,

SCIENCE TO THE RESCUE?

Establishing koa corridor in 1997. *Jack Jeffrey*

strapped on my pack, and dutifully stashed the keys to the Bronco in the gas tank compartment. Don and Alan had taught me that this was the best way to simultaneously ensure that you never lose the keys in the field and that all your coworkers can independently access the truck if necessary. On more than one nontrivial occasion, I came to appreciate the wisdom of this and many of their other field protocols.

I trudged through the tall grasses, relieved to have made it here in such good time and eager to finish collecting one last round of preliminary data and finalize my research plans. The morning dew was still so heavy that after only a few steps, my pants were thoroughly soaked from the tops of my rubber boots all the way up to my waist. But there was already some warmth from the sunshine slanting down through the trees, the ʻōhiʻa flowers were in full sparkling bloom, and a chorus of native birds was serenading me with their music. As if on cue, an ʻiʻiwi (*Vestiaria coccinea*) flew in out of nowhere and landed on a low-lying branch less than fifteen feet in front of me. One of the more common surviving members of the Hawaiian honeycreepers, adult ʻiʻiwi are bright vermilion with black wings and a long, curved, salmon-colored bill for extracting nectar (see Plate 5). They are such stunningly sweet and beautiful birds that despite all the other world-class candidates, ʻiʻiwi grace the covers of a large proportion of the books and magazines about Hawaiʻi.

I stood, mesmerized, as this ʻiʻiwi crept down the ʻōhiʻa limb, sipping nectar from its abundant flowers. Watching this graceful bird filled me with energy and optimism. I drank in the rich, moist air and felt grateful to be able to work in such places, follow my curiosity, and hopefully do some good.

That ʻiʻiwi had most likely flown in from a section of relatively intact, old-growth rain forest just below me. The first time I walked into that forest, I felt as if I was tunneling into a vast green sponge. Thick carpets of mosses, ferns, seedlings, and saplings blanketed every nook and cranny of its dripping understory, and layers of plants grew in, on, around, and through each other in the overstory. Best of all, everything was native—for once I was standing within a Hawaiian ecosystem with no weeds at all! The splendor and vitality of that forest left me speechless.

Several months later, however, I spent a day with Jack Jeffrey following one of his transects deep into another rain forest on the other side of the refuge. Over the course of that day, I saw how patchy this seemingly homogenous ecosystem could be: while some sections were relatively open, the vegetation in other areas was even more tangled and impressive than what I had observed below my field site. Was this patchiness representative of the primordial Hawaiian rain forest or an artifact of past human-caused disturbances? While this question may never be definitively answered, it was clear that feral ungulates were still significantly impacting this forest, as we saw copious amounts of scat, tracks, severely damaged native vegetation, and weeds. Jack explained that the mounded, tussock-type terrain we frequently encountered was also created by the past activities of cattle.

Unlike mainland tropical forests, where the greatest plant diversity is found in the tree canopies, most of the diversity in Hawaiian rain forests resides on or near the ground, perhaps because there has

SCIENCE TO THE RESCUE?

not been sufficient time on these young islands for a rich tree flora to evolve. Yet when we were in a particularly impressive patch of forest that day and I told Jack that I had never seen such a diverse and intact native understory in Hawai'i, he shook his head. "But you don't see all the plants that should be here but aren't because of the ungulates. For example, we should be seeing tons of lobelias and mints. And without the pigs and cattle, some of the plants we only see growing up in the trees epiphytically would actually be thriving on the ground."

Jack's observations were supported by his extensive experience in some of the most pristine rain forests remaining in the archipelago, particularly those in which some type of surrounding natural barrier had created long-term protection from ungulates and humans. I looked with new eyes at the layers of plants growing above us and thought of the pioneering ecologist Aldo Leopold's wry conclusion that the price of an ecological education is "living alone in a world of wounds."

Just above my field site was a forty-acre exclosure constructed in 1987 to provide a cattle-free arena for preliminary reforestation experiments. At that time this area still supported sections containing a mature koa-*ōhi'a* overstory, but most of the native understory had been destroyed and replaced by alien weeds. With the assistance of the US Forest Service, the refuge staff had tried to jumpstart koa regeneration within this exclosure by using bulldozers to experimentally scarify the soil beneath the mature koa trees and in the more open, treeless sections. Anecdotal evidence from other places in Hawai'i had suggested that running over and scraping the ground beneath koa canopies with heavy machinery could induce their hard, recalcitrant seeds in the soil seed bank to germinate, as well as stimulate the root systems of the established trees to send up new shoots. To my knowledge, however, no formal experiments had been performed, no rigorous data collected, and no scientific or even informal reports had been published. Jack explained:

> If we had seen significant koa regeneration in the open areas, we would have tried scarification on the rest of the refuge's degraded pastures because there were virtually no native plants out there except within the gulches. But we

didn't see any koa come up in the open. The seed bank was gone, maybe as a result of all those fires in the past. We did see tremendous regeneration under some of the mature koa trees, yet very little under others, and we didn't know why. But then some of the unscarified areas would also end up with about the same amount of regeneration as the scarified trees, so we decided mechanical scarification wasn't worth the time and money. I also found out later that scarification could shorten the lives of the big koa and ultimately even kill them. Over half of the scarified trees within that exclosure are now dead—you can go out there and find places where old mom has died and there are twenty- to thirty-foot trees growing everywhere underneath her.

The first time I explored that exclosure, the interplay between its native and alien floras reminded me of the ebb and flow of an intertidal marine ecosystem. In some areas, I found small islands of mature koa and *ōhiʻa* canopy trees overhanging a robust native understory surrounded by a vast sea of rank pasture grasses and impenetrable thickets of Florida blackberry. At first I was convinced that these scattered pockets of natives would, like so many ephemeral tide pools, inevitably be swamped by the mighty rising tide of alien species. But then I discovered some more interior areas where the native vegetation seemed more like the rocks and dunes beyond the reach of the highest tides, and there were even a few extraordinary places where it appeared that the *aliens* might eventually be drowned by a sea of natives. Which way was this ecosystem headed? What would happen if we did nothing except maintain the fences and keep the ungulates out? As I searched for answers to these and a suite of related questions, I found a dearth of hard data and a surplus of confident yet often contrasting personal opinions.

I eventually discovered that a few colleagues in the Forest Service had previously investigated the patterns of native recruitment in that exclosure. Over the course of this research program, Don and Alan had spent countless hours exhaustively searching for naturally establishing native species. One of their most interesting results was that although downed and decaying koa and *ōhiʻa* logs comprised less than 2 percent of the potential seedbed area, about 70 percent of the more than two thousand regenerating native individuals they found were rooted in those logs. Several of these species also grew more rapidly within the logs and other types of organic substrates, such as moss-covered root mats and pockets of organic matter within the crevices of live trees.

These results were further supported by more casual observations of native plant regeneration throughout the refuge. Before the refuge staff began their restoration efforts, about the only place one could find native regeneration in the high-elevation pastures was on top of giant fallen koa and 'ōhi'a logs. Moreover, even after the pigs and cattle were removed and a native seedling no longer needed to be beyond the reach of a grazing animal to survive, the vast majority of native recruitment continued to occur on these decaying logs. Regeneration also began to happen on the smaller logs that had previously been accessible to the cattle. In some cases, the plants emerging from these logs grew with amazing vigor because they were actually already several years old; over the years in which their vegetation had been repeatedly browsed back, they apparently had developed extensive root systems capable of supporting rapid above-ground growth.

Virtually every time I spotted native plants emerging out of the pasture grasses, I found them to be growing on old fallen logs, although in some cases the grasses had grown so tall and the logs had decayed so much that I would never have detected this phenomenon if I hadn't been looking for it. Yet in contrast to the regenerating natives clinging to those decaying logs like lifeboats, the grasses and other weeds tended to be rooting almost everywhere *except* the logs. These observations led me to wonder whether it might be possible to facilitate the restoration of the refuge's pastures by manipulating their substrates so that they favored the "natural" establishment of native species.

I eventually came up with my "Nurse Log Experiment" (although others persisted in calling it the "Log Cabin" experiment, even though my last name is pronounced "Kay-Ben") one day when I hacked my way through a dense blackberry thicket to inspect another decaying log. When I got there, I discovered that although that log was engulfed by the blackberry, none of its canes had penetrated the log itself, and when I peeled them back I found that the log's upper surfaces were covered with native seedlings.

I knew the establishment of plants on these so-called nurse logs was not unique to this refuge because I had previously observed native plants growing on fallen logs in other Hawaiian forests and in the temperate rain forests of the Pacific Northwest. I also knew this was probably not just a recent, human-caused phenomenon because some of the old-growth trees within these ecosystems had big curved holes where their massive trunks reached the ground. The shape of these missing sections strongly suggested that those trees had established on nurse logs that had long since rotted away.

As usual, these observations led to a cascade of seemingly endless questions and hypotheses as I thought about how to design a nurse log experiment at Hakalau: Why were those decaying logs so advantageous for native plant establishment? Did the plants growing on them experience less competition than the plants growing on the ground? Did the logs provide more nutrients and water? Better drainage? Was there more available sunlight up there or more protection from frosts? Could there be a distinct and more beneficial mycorrhizal fungal community growing within the nurse logs (the great majority of plants within and beyond the Hawaiian Islands require these fungi to germinate and grow)?

Or might it be something as simple as being up off the ground reducing the odds of getting crushed? A senior botanist from Hawai'i Volcanoes National Park had recently told me how she and a colleague once placed "seedlings" made of plastic drinking straws into a variety of niches within a native rain forest. At the end of their experiment, they found that the straws on the ground were much more likely to have been destroyed by falling organic debris than were the straws up on the logs, tree ferns, and live trees.

Or maybe the term "nurse logs" was a misnomer: perhaps they did not actually provide an advantageous microenvironment at all. Maybe Hawai'i's prehuman rain forests were so crowded with fallen logs and tree ferns that there simply was little exposed soil available, so that over evolutionary time natural selection favored plants that were able to *tolerate* growing on those fallen logs. This hypothesis might also explain why the alien species did not appear to utilize the nurse logs like the natives did. Then again, most of Hakalau's alien species were herbaceous, while most of the native species growing on the logs were woody trees and shrubs. Or maybe it had something to do with different seed dispersal syndromes or that all the native seeds that landed on the ground were eaten by exotic seed predators?

As these and too many other ideas rattled around in my brain, I thought of a plenary presentation I had attended many years earlier that had been given by an eminent Amazon rain forest ecologist. Although his research program had been extremely productive, he appeared to be similarly overwhelmed by the number of important questions he knew he would never be able to even investigate, let alone answer. He concluded his talk by showing a cartoon of an eager dog standing on the edge of a vast forest preparing to urinate. The caption read, "So many trees, so little time."

My first experience conducting a real scientific research project came near the end of my undergraduate career when I spent a summer studying a rare milkweed in Vermont. In my naivety, I frantically quantified and recorded everything I could think of for each and every milkweed plant I encountered. I later learned in graduate school that any idiot can go out and collect massive quantities of ultimately useless data. Even though acquiring this information typically requires long, tedious hours of physically demanding labor, most of us relish this work because, as one of my colleagues succinctly puts it, "a bad day in the field beats a good day in the office." Similarly, many of us love to get together (ideally after a few beers) and hold forth on how these data support our pet theories that cleverly explain the endlessly intriguing patterns and processes we observe in nature.

Yet field biologists who merely collect data and pontificate about it are like garage musicians who never record or publicly perform their music: while they may be highly skilled and dedicated, few outside their immediate circles will ever be able to appreciate and build upon their work. However, because most formal research programs require substantial amounts of public resources, scientists tend to be under much greater pressure than musicians to broadly distribute the fruits of their labors.

Few nonscientists appreciate how much skill and perseverance are typically required to produce a rigorous scientific publication. And because good science journals always receive far more papers than they can print, even excellent research must often be rejected. The path to publication is usually far more difficult for people working in nonresearch fields such as education, resource management, and conservation. This is because even if they have the ability to perform and write up high-quality research, they rarely have much incentive to do so because such work tends to mean little to their colleagues and supervisors. Moreover, getting tomorrow's lecture ready or dealing with the latest shopping mall proposal must understandably come first. One of the cruel ironies of all this is that while the ecology of our planet remains largely unexplored, most of the field data we so laboriously collect and analyze never makes it into print.

When I first began working in Hawai'i, I was flabbergasted by the dearth of even basic ecological publications. While some would confidently tell me about the habitat requirements of a rare native species or the best way to control an invasive weed, I soon discovered that virtually all such claims were supported only by personal convictions and unpublished anecdotal observations. I also frequently encountered at least an implicit belief among long-term residents that ecology and conservation just worked differently in Hawai'i. The authors of the groundbreaking 1987 *Manual of the Flowering Plants of Hawai'i* made a similar observation in their introductory remarks:

> A certain amount of mysticism had crept into Hawaiian botany over the years—including the notion that a period of initiation was necessary before a novice botanist could be admitted into professional practice! This was really not true, but it made access to Hawaiian botany difficult for nonresident workers. There also seemed to be an implication that evolution worked differently in Hawai'i than anywhere else on earth, and that variation seen in Hawai'i was somehow more significant than similar variation found elsewhere. . . . There probably is no area in the world for which the classification of flowering plants has been more confused in recent times than in Hawai'i.

Warren Wagner, lead author of the *Manual,* told me stories of how some prominent Hawaiian botanists would hoard their specimens and refuse to let outsiders even look at them, much less review and critique their taxonomic work. "Some of these guys were so taken with the morphological variation they observed in the field that they would end up doing things like classifying different branches from the same tree as different species!" He explained that one of his overarching goals in writing the *Manual* was to replace this kind of sloppy, inconsistent, and parochial botany with one grounded in the rigors of modern science. "We wanted to clear the dust and create a foundation for further work in systematics and conservation—we had to know what was out there to be more successful. If there was to be any hope to save the Hawaiian flora, we knew we were going to need this deeper understanding of the species and their lineages, especially for restoration projects."

In many respects, the publication of the *Manual* fulfilled these goals admirably. As the authors noted in the preface of their revised 1999 edition,

> Because of the availability of the *Manual,* field work during the 1990s has been more focused. . . . Over 180 additional naturalized species have been documented. To these have been added 233 new distributional records of species already established in the islands. Thirty-three new taxa of native plants have been discovered. . . . Field studies have resulted in the rediscovery of 27 species presumed extinct, the resurrection of 25 taxa relegated to synonymy in the *Manual,* and more precise information on the condition of much of the endemic and indigenous flora. When the manuscript for the *Manual* was submitted for publication, 19 Hawaiian plants had been federally listed as either threatened or endangered; that number is now 271.

I similarly wanted to help put the largely trial-and-error, idiosyncratic world of ecological restoration and conservation biology in Hawai'i on the more solid scientific foundation of modern ecology. Yet because I was deeply involved in the islands' often desperate conservation battles, it was equally important to me to find a way to make my "real science" relevant and valuable to the "real world." Thus after much struggle, I finally settled on two overarching goals for my Hakalau research program: (1) Provide concrete, feasible tools and ideas to help preserve and restore native species within and beyond the refuge; and (2) produce high-quality research publications that significantly contribute to our more basic scientific knowledge.

The general question I chose to investigate was whether the conditions favoring the germination and establishment of native rain forest species might be different than those favoring alien species. As the intellectual framework for my Nurse Log Experiment slowly came together, I began spending more time thinking about how best to translate my ideas into actual experiments. I knew from hard experience that even the most brilliant scientific questions can be ingloriously torpedoed by poor methodological decisions and mundane logistical issues. For example, exactly how should I define and standardize a "decaying log," where would I get these logs, and how would I transport them to my field site? Which native and alien plant species should I seed into my experiments, and how should I collect, process, and sow these seeds?

When professors begin working in a novel experimental arena, they often foist the job of solving these kinds of practical problems onto their graduate students, research technicians, and postdoctoral fellows. While I did not have any such people at my disposal, I was entitled to a portion of Don and Alan's time. I felt exceedingly lucky to have their help, because both men were extremely knowledgeable about Hawaiian ecosystems in general and the Hakalau Forest Refuge in particular since they had been working up there for decades. And in spite of their constant gaiety and slapstick humor, they were two of the most skilled, conscientious, and hardworking field technicians I had ever seen.

Unlike the other scientists and scientists-in-training I was used to collaborating with, however, Don and Alan had no professional stake in the outcome of this or any other research program. Indeed, the whole paradigm of using reductionist experiments to tease apart complex ecological phenomena such as native versus alien plant regeneration was foreign to their more holistic worldviews and intuitive, commonsense approaches to resource management problems. They had also both grown up under the "plantation mentality," in which locals were expected to keep their mouths shut and do what they were told. Thus they would go to great lengths to avoid telling me or any other professional white person that we were wrong or that they knew a better way to do something. Yet somehow they still seemed to receive the lion's share of blame when things went awry. For example, Alan once told me how a team of mainland Forest Service scientists got so frustrated by the inconsistencies in their field data that they instituted a policy in which every line of data had to include identifying initials.

"Well," I asked him, "How did that policy work out?"

"Pretty well . . . for about a week. Then someone started analyzing it and found that all the mistakes were coming from the scientists' data sheets!" Alan closed his eyes and beamed his enormous, beatific smile. "For some reason they abruptly dropped that initializing policy and went back to yelling at us again!"

When I received a letter from the Hakalau Forest Refuge in the spring of 1998 granting me permission to conduct my proposed Nurse Log Experiment, I took a break from my grinding office work to fantasize about all the great things that would happen because of this research. I pictured teams of technicians assembling thousands of artificial nurse logs to air-drop into degraded areas at Hakalau, the other Hawaiian Islands, and then throughout the world. Once on the ground, these logs would catalyze powerful cycles of native species establishment and alien species suppression in ever-expanding circles until all of the earth's ecosystems would be effectively restored back to their original shining glory. Of course, all the textbooks would have to be rewritten to incorporate the new ecological paradigms and applied management strategies created by my Nurse Log Research Program. Scientists and resource managers would attend harmonious meetings together, hold hands, and laugh incredulously about the animosity that used to exist between these two professions.

4   **LAULIMA**

Five months later, I drove back up to Hakalau to recensus my experimental nurse logs. I was hoping to find at least a few new seedlings among the 2,970 different spots into which we had sown a seed or transplanted an ʻōhiʻa seedling, but as I walked from one barren plot to the next I found that little had changed since my previous visit one month earlier. Although I knew that many Native Hawaiian species were notoriously slow to germinate and establish even under ideal greenhouse conditions, I couldn't help wondering why so few of my seeds had sprouted. Could it be the extended drought we were in? Exotic seed predators such as mice (I frequently saw what looked like mouse tracks running across some of my treatments)? Could the seeds we harvested have been immature or infertile, as had been the case in some of my experiments in other Hawaiian ecosystems?

As I walked through a thick tangle of blackberry that led to my last plots, I told myself to quit worrying about everything: I had designed a good experiment, and Don and Alan had done their usual superlative implementation job, so there was nothing to do now except patiently wait and see.

I smiled at the memory of the day we had launched this experiment. Having learned how hard it could be to simply walk through this terrain, I had organized a large work crew to help with what I knew was going to be a lot of physically demanding tasks. Of course Don and Alan had turned the whole operation into a party, but somehow, amidst all the good food, storytelling, and laughter, we managed to get far more done than I had dared hope was possible.

Several years later, while we were sharing a few beers after a long day in the field, Don told me the Forest Service merit award he had received at my recommendation for his work on this project was the only complimentary recognition he had ever earned throughout his entire professional career. "And your getting all those students and interns to help us set up that field experiment was the first time anyone ever thought about us and our needs," Alan added. "You know how some of the scientists used to treat us? At the end of the day, after we'd gotten all their stuff done, they'd put their datasheets away, and then they'd put their equipment away, and then they'd put us away."

"But sometimes they'd forget and leave us out in the rain," Don smirked. "That's why we're so rusty and creaky now!"

After one last handful of ʻakala berries, I packed up my equipment and headed back up the mountain. When I reached the top, I parked next to the refuge cabin, grabbed my clipboard and datasheets, and ducked inside the plastic curtain that served as their greenhouse's front door. As usual, I found Baron Horiuchi, Hakalau's horticulturalist, up to his elbows in plants and dirt.

"Howzit, Bob!" he said with his big smile.

"Hey, Baron. What are you up to today?"

"ʻŌhiʻa, ʻōhiʻa, and more ʻōhiʻa!"

"From seed?"

"Some seed, some snatchlings," he laughed, pointing to a tray of healthy-looking transplants. Baron had previously told me that if he needed a bunch of ʻōhiʻa and didn't have time to wait the two years it could take to germinate and grow them from seed, he sometimes "snatched" some wild seedlings. At first I wondered whether this technique might be a case of robbing Paul to pay Peter, but I came to appreciate the benevolence of his approach after I saw the thousands upon thousands of 'ōhi'a seedlings that carpet the understory of some of the refuge's more pristine old-growth forests.

But because there were no carpets of wild seedlings of all the other native species the refuge needed for their restoration program, Baron had to figure out how to grow them all himself. "When I first started," he told me one day when I was picking his brain for ideas for my own experiments, "I tried to look for basic information in books, but a lot of these species had never been propagated. So I decided to just look at what nature was doing and try to imitate that, and for

Baron Horiuchi in the Hakalau greenhouse. *Jack Jeffrey*

the most part that worked out pretty well. But then I started working with the endangered plants—there are so few of them, and there is no regeneration occurring in the field—so I would go out into the forest and try to find and collect some ripe fruits before the rats got them, then take off all the pulp and let the clean seeds mature. Then I just tried lots of different things and waited to see where germination was best. It was fun, but it took a lot of time, and the results were often completely different for each species."

I walked over to another section of the greenhouse to look at Baron's collection of rare and endangered species that would eventually be transplanted back into the field. I knew that many of these plants were literally on the edge of extinction—some were down to just a couple of individuals remaining in the wild, and a few had even been presumed extinct before someone stumbled into one while traipsing around the refuge.

Looking at their exotic beauty and contemplating their vulnerability always gave me chicken skin. As was the norm in the other native plant greenhouses I had visited across the state, some of these were the last specimens of an entire *species,* yet here they were, just sitting on these crude benches in the middle of nowhere. Any one of a number of not so implausible events—a hurricane, a cow, a fungal outbreak, a human vandal—could result in their permanent extinction. Yet unlike rare stones or works of art, none of these plants was protected by bulletproof cases, closed-circuit video cameras, or security guards.

"So how are your porcupines doing?" Baron asked, jolting me out of my endangered species nightmare.

"That's just what I've come to find out."

We walked over to the only section of the greenhouse that was not overflowing with plants. Baron called this my "porcupine experiment" because most of the treatments still largely consisted of dense arrays of protruding toothpicks marking the location of small seedlings and seeds that had yet to germinate. My quick visual survey told me that while there had been some new germination since my last visit, the overall fertility of this experiment was still disappointingly low. Although Baron and the rest of the refuge staff continued to be exceedingly supportive of my research, I couldn't help feeling guilty about usurping several of their precious benches for what might well turn out to be a failed or at least unhelpful experiment.

Fortunately, Hakalau's restoration program was not dependent on the results of my or any other formal scientific research program. Jack and Baron were also not the kind of people to use our collective ecological ignorance and inexperience as excuses for inaction. On the contrary, I knew that both men relied primarily on their own intuitions, observations, and informal experimentation to tackle the refuge's endless series of applied problems.

In fact, Jack told me many stories of scientists who after a few years of research argued that the refuge should implement some narrowly focused management strategy or follow a particular academic theory. Although he was quite appreciative of formal science in general and the external research performed at Hakalau in particular, he had grown increasingly skeptical of the practical value of this work. "When the final scientific papers and reports come back to us," he explained, "they are almost always focused on one or a few species. Often they are quite interesting, but because we are managing the refuge for the ecosystem as a whole, they're worthless as far as management recommendations. And the theoretical stuff is real nice on a small scale, but as soon as you go to a larger scale . . ."—he shook his head. "For us, it has just been a lot of seat-of-the-pants experimentation and learning as we go."

Thus in much the same way as I had formulated my own Hakalau research program (albeit with much more extensive firsthand knowledge of and experience in this ecosystem), Jack and his colleagues created an initial refuge restoration program based largely on what they had seen and discussed with others. Unlike me, however, their program was not designed to answer any specific questions or employ the kinds of standardized methodologies necessary for rigorous science. Indeed, although carefully researched and discussed, from the beginning their plan explicitly employed a philosophy of "let's see what works, what doesn't, and revise accordingly."

While their overarching goal was obviously to reforest the refuge's vast upper-elevation degraded pastures and reconnect them to the relatively intact lower-elevation rain forests, the best way to accomplish this goal had been anything but obvious. As Jack recalled,

We wanted to create higher elevation, high-quality habitat for the native birds as quickly as possible [if global warming results in mosquitoes surviving at higher elevations, the birds will need to keep moving upward to escape avian malaria], but we knew that the birds wouldn't leave the existing intact forest and fly out over those open, treeless pastures. I'd read some papers about the efficiency of planting groves, but we decided against it because we knew it would take far too long to plant little blocks here and there—if we had used that approach, we wouldn't be very far across the refuge today. However, we saw that the birds would fly up along those long tree-lined gulches. So we thought why not plant in long corridors—it's logistically much easier than groves, plus the birds might utilize them and drop native seeds along the way and maybe we'd get some natural regeneration, which would really help us because there's just no way we will ever be able to replant all those pastures.

In 1989, after they finally got the first subunits of the refuge fenced and the ungulates out, the staff and a band of dedicated volunteers boldly began planting long strips of koa trees up and down the pastures. They hoped that over time, the emerging native forest might "naturally" begin to fill in between the corridors and shade out the grasses. However, since nothing at this scale had ever been attempted in Hawai'i, they continually had to invent and refine their methodologies as their program progressed.

For instance, when they first got started, the standard protocol for planting koa trees was to grow them in big pots, carry them out into the pastures, dig holes in the grass with a shovel, and then painstakingly transplant the trees into these holes. But when they found that people could only plant a few trees each day by following this protocol, they began looking for better and faster ways. They eventually discovered that if they grew the koa in long, narrow dibble tubes instead of pots and ran a bulldozer mounted with a sod-scraping, three-pronged rake across the planting areas, their crews could plant over two thousand trees a day. "This gave us a path to walk on and follow, too," Jack explained, "which is essential when you've got people stumbling around through all that tall grass out there. The scraping also kept the grass away for a year or two, which proved to be enough to get the trees established. So we just followed the path of least resistance up and down the pastures, and the volunteers didn't even have to think—just follow the line and plant!"

By employing this simple but highly efficient methodology, by the end of 1989 the staff and their steadily growing volunteer corps had planted thousands of koa seedlings in corridors containing three rows of trees spaced about twelve feet apart. However, they later discovered that in some areas the subsequent growth and survival of those koa were poor. After extensive observation, monitoring, and analysis, they concluded that frost was killing or at least weakening most of the koa trees that failed to establish and thrive. Once again, they immediately began experimenting with a wide variety of methods to solve this problem. Ultimately, they tried over a dozen different

Establishing koa corridor in 1997. *Jack Jeffrey*

frost-protection devices, including mulching with rocks (to absorb heat during the day and then release it back to the trees during the cold nights) or bark (to release heat as it decomposes), the Wall-O-Water (a commercial product used to protect tomatoes), and some mini-greenhouse contraptions.

Jack also had a hunch that the cold winds that frequently blow across the pastures might be adversely affecting the koa seedlings. He recalled,

> In the old days, before we had any bathrooms on the refuge, you'd get up at 4 a.m. to do the bird surveys, and if you had to go you'd grab a shovel and walk out into the pasture. But if you weren't quick enough, you could get a little frostbite on your butt cheeks from the wind blowing down the mountain. So I thought it might just be that those early morning winds were enough to push the koa over the edge and kill them. I knew if we put up something like plastic it would eventually get blown away or destroyed, so we tried black shade cloth suspended between two wooden stakes *mauka* [the upper or mountain side] of the trees to hopefully deflect just enough of the wind to keep them alive. Then I thought, well, since this is an experiment, we also better put some up *makai* [the lower or ocean side], too.

When they recensused their experiments six months later, to their great surprise they found that about 90 percent of the koa protected by the *mauka* shade cloth treatment were dead, but every single tree with a *makai* or east shade cloth had survived. While a few of the other frost-protection devices had also worked well, they decided these treatments were too expensive and labor intensive to effectively employ on a large scale. But since their shade cloth structures were essentially two sticks in the ground, they conducted a follow-up experiment to compare the side-by-side performance of the *mauka* versus *makai* protection treatments. When they again found 100 percent *makai*-side survival versus about 75 percent *mauka*-side mortality, even though they did not understand why they worked so well, they concluded that the *makai* shade cloth structures were the way to go.

Hakalau's koa seedling "frost protection" device.
Jack Jeffrey

A team of scientists eventually concluded that those outplanted koa seedlings apparently could survive a moderate amount of frost, but if they defrosted too quickly their cell walls could rupture and desiccate. Thus it appeared that the *makai* shade cloth treatment was saving the koa in part by blocking the morning sun and slowing down this warming process. They also found that these shade cloth structures helped protect them by blocking out about half of their exposure to the cold night sky and keeping them one or two degrees warmer by reflecting some heat off the ground. In addition, they may have increased the amount of water available to the koa seedlings by providing a condensation surface for the refuge's often abundant fog.

Encouraged by their success, Jack and his crew just kept planting, observing, experimenting, and refining their protocols as necessary. They ultimately discovered that if they planted and fertilized the koa seedlings in the spring or early summer, by the time the first frosts hit in November, most of the trees in the low- to mid-elevation regions would be above the one-meter-tall frost-kill zone and therefore would not require the shade cloth protection. However, they found that although the flatter sections of the refuge had good, deep soils that supported the fastest plant growth rates, cold, frost-producing air tended to flow down to and get stuck in these areas, and thus in some cases virtually all of these plantings died despite their best efforts to save them.

While they continued to mostly plant large numbers of the fast-growing, hardy koa, they also began to grow and plant out other native species as time, labor, seed availability, and greenhouse space allowed. In 1996, the refuge finally hired Baron Horiuchi as their official horticulturalist. (The US Fish and Wildlife Service technically does not have "horticulturalists," but the refuge staff modified a horticultural job description from the National Park Service that somehow appeased their bureaucracy. Baron told me that as far as he knew, he was the only "horticulturalist" in the entire FWS.) In addition to growing some of the native rain forest plants other than koa, including many endangered species, he spent a great deal of time working with the refuge's ever-growing number of volunteers.

Baron was born on Oʻahu but moved to Hilo when he was one year old and never left. One day I asked him about his awareness of environmental and conservation issues while growing up on the island, and how, as the son of a fisherman, he ended up propagating plants at Hakalau. He told me,

> I'm one of the lucky ones. I grew up with people who didn't even know what a native species was, but eventually I got interested in horticulture and studied plant propagation in the College of Agriculture at UH–Hilo. I got to work with a

fellow named Donovan Goo, who taught me a lot about how to grow native plants, and he took me up to Hakalau. At that time we were just working in the pastures, and I could actually see and feel the land hurting, and how much it needed to be healed, so I decided that that was what I wanted to do. Obviously, it soon became more than a job for me. It's a way of giving back and taking care of the island. And I guess since I don't have kids, the plants have become my kids. When I walk through the native rain forest now, I think about what happened in the past, what the trees were used for, how the people used to take care of and live off the land. I have a friend who is pure Hawaiian, and even though I'm Japanese he tells his friends that I'm more Hawaiian than most of the Hawaiians because I'm helping to take care of the islands. And when I take my friends up to the refuge, they're always so impressed! Tears come to their eyes when they walk into the forest and see the plants and hear the birds.

When Baron first started working at Hakalau, the refuge's "greenhouse" was just a simple metal frame. Since there was little funding for his horticultural operation, Baron began by finishing up the greenhouse himself, which included making homemade benches and jury-rigging an irrigation system. Because there is no county water up there, he installed two tanks on a hill above the greenhouse. However, this system generated only about fifteen pounds per square inch of water pressure.

"All the experts told me you can't run overhead irrigation with fifteen psi, so they urged me to use drip irrigation instead," Baron recalled with a laugh. Then he explained:

> But I knew we'd never have enough plants to do what we needed to with a drip system, so I started experimenting and found that there was enough pressure to run a sequential, bench-by-bench overhead system. I set it up so that every two benches had a valve and a timer, but all we could afford was really cheap hardware. The guys that make that stuff

have it down perfect—as soon as their warranty expires their things break like clockwork! But then one day a volunteer said her family wanted to make a contribution to the greenhouse. She told me to go look for some high-quality, dependable irrigation equipment. So I called her up and said, "Are you sitting down? The stuff we really need is going to cost about $1,500!"

But she said, "We can swing that. Anything else you need?"

"Well, there's a solar power converter that goes with it for $300 more." Within two weeks we received a $2,000 check. That equipment helped us tremendously, and it's still in operation today.

By 1997, their operation was humming along so well that Baron was able to grow about five thousand non-koa seedlings, plus another twenty thousand koa. Yet again, Baron and Jack relied on their own experiments to develop and refine their methodologies for planting these relatively slow growing, less hardy non-koa species. Two of their biggest challenges were the frosts and the droughts. During El Niño years, the refuge might not get a drop of rain for three months, then over thirty inches of rain in one weekend, and then another three months with nothing. Commenting on their results, Jack said,

> I expected 'ōhi'a to do fine out there in the pastures, because it's a primary succession species. Same thing with the other native understory trees and shrubs—we felt sure most of them were going to do fine. But in the end, nothing survived in the open pastures—zero! Yet we had 80–90 percent survival when we planted within our koa corridors. I can't say it was the reduced competition from the grasses because the grass cover wasn't much different beneath the trees than it was adjacent to the corridors, so maybe it was just the frost protection. At any rate, we eventually figured out that as long as we had at least 25 percent overhead cover of koa canopy, the survival of our non-koa species was good, so from that point on we just planted everything within the corridors.

Thus koa turned out to be the "forest engineer." Without the protection provided by their canopies, it was virtually impossible to establish the other native forest plants in all those exposed pastures, but by utilizing their ever-expanding corridors of koa, real rain forest restoration suddenly seemed very possible.

In about a dozen years, they managed to establish koa corridors (spaced 100–200 yards apart depending on the topography) all the way across the pastures and up to their fence at the top of the refuge. By 2011, the corridors stretched across the entire refuge and included over half a million native trees and thousands of endangered plants.

"We did have some gaps within the corridors due to mortality," Jack recalled. "We had six years of cold El Niño weather, so we stayed away from some of the higher elevation sites for a while because we couldn't make enough frost-protection devices. We also increased our planting distance between the koa from twelve to twenty feet because

Establishing koa corridors in 1997. *Jack Jeffrey*

we found they closed canopy so fast. Maybe even twenty feet is too close, but my feeling is that if down the road somebody comes in and says 'boy, they shouldn't have planted so many trees,' they can get out their chainsaws and do some thinning."

When I asked the refuge staff how they have managed to implement such a large and successful restoration program, the first thing everyone said was "the volunteers." They explained that with their tiny staff, modest operating budget, and myriad other responsibilities, there was simply no way they could ever have even attempted such an ambitious restoration project on their own. Each staff member also stressed that in addition to their labor, the refuge's volunteer program generated a tremendous amount of public education and outreach that in turn supported their restoration program in many direct and indirect ways. Jack added,

> It also takes a lot of leadership and love for what you're doing to inspire the volunteers to do this kind of work. Some of them are already very dedicated before they even get here, and thus you can use and abuse them and they keep coming back for more. But we also get lots of people who are not used to being out in the field. For example, lots of office workers from Honolulu, who pay their own way over to the Big Island and have to get up at 4 a.m. on Oʻahu to catch a flight to make it here on time. So we learned that rather than work them until 6 p.m. in the rain and cold on their first day, it's far more effective to start more gently, take them birding at first, give them a Hakalau Forest T-shirt. Once we started doing those kinds of things, our volunteer applications skyrocketed. Now we are getting everyone from hardcore wilderness freaks and CEOs of major corporations to mainland school groups and environmental educators. Many of these people would be staying in five-star resorts if they were vacationing in Hawaiʻi on their own, yet they come here and say, "You now have my life for a week, what challenge are you going to give me?" They end up working their butts off for us, yet when they leave they always tell us how wonderful it was, and we get lots of pats on the back.

It's really hard to tell how our volunteer experience affects everyone, especially the kids. Most of the older folks are already in the choir—that's why they are here. But when we take kids from the city who have never been exposed to anything, bring them out to the refuge to plant trees, is that going to turn them around? I'm never sure . . . maybe it will for some. Many are at least open-minded and impressionable, although some of them *really* don't have a clue—sometimes they don't even know what island they're on! But there are also kids who don't seem to be engaged at all, yet later you'll hear them talking to each other and realize just how much they did absorb and how proud they are about what they did. Over the years we have managed to lightly mentor a few who we hope might one day end up being our replacements.

For Baron, working with the volunteers has also been much more than just a way to get lots of plants in the ground. "Sometimes I'm just amazed by how strong human spirits are," he told me, shaking his head. He continued,

> Our volunteers just really want to give. They work eight to ten hours per day, in difficult terrain under difficult conditions, and yet you can just feel how hard they're working. And we become so close working together. I have a hard time putting all this into words. . . . I tell them at Hakalau there are only good feelings and spirits because all these good people come up and help and leave all this good energy behind.
>
> Of course, planting trees is a way of giving back and making sure in the future there will be Hawaiian forests and hopefully Hawaiian wildlife. The feeling of hands in the dirt, planting in lands that were once forest, seeing plants that you rarely see becoming more abundant because of all of our efforts. But there are a lot of other things involved too, like spirits. For instance, there was this lady who planted koa up on the refuge way back in 1987. She came back a few

years ago and I took her to her trees and she went over and hugged them and then just broke down and cried. Anyone who knows her knows she's a very strong-willed person—she doesn't do stuff like that!

When we get back up to the cabin after planting all day, everyone is always really tired, dirty, and stinky. But then I tell them we have some large, seven-gallon pots in the greenhouse with special plants that are about three, four feet tall, and if they want they can select one and plant it as their personal plant. Despite how exhausted they are, pretty much all the volunteers go right back out and do a really good planting. They'll lay down with their plant, talk to it, give it some water. Last year a couple of girls sang songs to their plants, then they wanted to sleep with them! Most will also

Kamehameha Schools volunteers working at the Hakalau greenhouse.
Jack Jeffrey

write something on the aluminum tags I give them. I don't know what they write because they're private and I never read them. The only real problem with this system is that people are always asking me, "How's my plant doing?" There are so many plants out there now I have a hard time keeping track of them all, so I tell them they have to come back and tell me!

Finally, I asked Baron to tell me the story behind the beautiful "*Laulima*" sign that hangs in front of the refuge's greenhouse.

In this context, *laulima* means "many hands working together." Back when we were getting the greenhouse going, one of my first volunteer groups was from Honoka'a High School. At one point all of our hands were in the wheelbarrow, and one of the kids said, "Wow, look, many hands working together!" I thought that was pretty interesting, so I called up Don Goo and asked him if there was a Hawaiian word for that, and he said, "*laulima*—why you wanna know?" I told him the story and said I wanted to make a sign and dedicate the greenhouse to the volunteers.

The next day someone at DOFAW [the state's Division of Forestry and Wildlife] said, "Hey, we've got a letter-making machine, we'll make the letters for you." So now the sign itself was *laulima*—Forest Service, DOFAW, and USFWS. Then later I worked with some students from Hawai'i Community College for a few years. When they graduated, they gave me the *laulima* sign that's up there now. They made it out of *kamani* wood [*Calophyllum inophyllum*, a member of the mangosteen family brought over by the Polynesians], with imprints of hands on it and a little hole in the outline of the Big Island to show where the refuge is. There's a lot of meaning in that sign for me.

5 PLACE OF MANY NEW PERCHES AND FEWER HOOVES

Over the next few years, Don, Alan, several Native Hawaiian Forest Service interns, students, volunteers, colleagues, and I collectively spent many hundreds of hours working on that Nurse Log Experiment. In retrospect, I believe we learned a tremendous amount from performing that research, talking amongst ourselves and with the many other people we encountered along the way, and perhaps most importantly, simply spending time together in the field.

In the end, however, luck was not on our side. Perhaps due in part to the extended drought that occurred throughout much of that experiment, relatively few of our seeds ever germinated in the field, and few of the seedlings that did emerge survived to the end of the study. We did eventually get more seed germination and seedling establishment in the greenhouse, but no clear overall patterns emerged. Some native species did better under one germination substrate/light combination and poorly under another, while the reverse was true for other native species. There was also substantial intratreatment variability across the field and greenhouse replicates. For instance, seed germination of several species was relatively high on half of the decaying koa logs in the shade but low within the remaining replicates of this same treatment combination. In other words, as a whole, this research turned out to be an unpublishable mess.

I also realized in hindsight that I had made several mistakes that seemed to have become all too common in the worlds of conservation biology and resource management in general and within the Hawaiian Islands in particular. First, I investigated too many variables that were too difficult to standardize and control. For example, due to the often extreme ecological heterogeneity we encountered on the refuge, one "decaying koa log" did not necessarily turn out to be the same as the next. Second, at Hakalau and several of my other research projects, I became increasingly disconnected from the nuances of my own experiments as my role gradually shifted from field ecologist/restoration practitioner to manager/bureaucrat. Finally, I slowly sank deeper and deeper into the quicksand of workaholism. Like so many of my colleagues, I was not good at saying no, and I took on far more than I could handle. Trying to juggle all my projects and mushrooming responsibilities while coping with the formidable federal bureaucracy ultimately turned me into one overextended, frazzled, and burnt-out employee.

I still remember the day when, after yet another insanely long and stressful week in the trenches, I decided for once to try and enter the actual hours I had worked during that pay period. I had some vague delusion about earning some comp time and taking a little vacation the following month, but of course the Forest Service's arcane software refused to accept my unconventional data (starting the workday at 4 a.m. did not compute). Just then, Alan happened to walk by and heard me cursing.

"Are you fighting with the computer again?" he asked, peering over my shoulder.

"This stupid program won't accept my hours!" I snapped. "You got any suggestions that don't involve sledge hammers?" Alan was proud of his computer incompetence—he would choose to run transects through gorse over office work any day—but in my desperation, I hoped that somehow even he might have learned a few tricks over the years.

"Well, let me see . . . hmmm. . . . You got lots of hours there, Bob," he said, shaking his head. "Yeah, actually, there is a special code Don and I developed for that."

"Really? What is it, because I want to get the hell out of here!"

"Try entering One-Dee-Ten-Tee."

"Huh?"

"One-Dee-Ten-Tee," he repeated. "Write it down using as few letters as possible. I assure you, it applies perfectly to your situation."

"Thanks Alan, I'll give it a try."

"No problem, I'm always happy to help you out with your computer problems." He laughed heartily and left.

After failing to get their code to work, I asked Alan the following week to write it out for me, but he just shook his head emphatically. While Don cackled in the background, Alan earnestly explained that the only way to really appreciate their code was for me to figure it out myself. But after failing yet again at the end of the next pay period, I'd had enough.

"OK, I've tried everything," I said bright and early the following Monday morning, "and I'm getting that sinking feeling you're putting me on again."

"No," Alan deadpanned, "I assure you this is a really important code for people who work the kind of hours you do. Don and I secretly developed it years ago for one of your predecessors, but I'm telling you about it now because I think you may be the first scientist capable of grasping its significance."

"Do you guys ever use it yourself?"

"Not us," Don exclaimed. "We avoid it like the plague!"

"Yeah," Alan agreed, "we always seem to get this mysterious illness that prevents us from working enough hours to use that code."

"Then let's get to the bottom of this once and for all!" I sat down with pencil and paper and glared at them.

"I told you," Alan said, "write it out using as few letters as possible."

"One-D-Ten-T."

"You're getting closer," Don said. "Just use numbers instead of letters where you can."

"Shush!" Alan scolded. "You're giving it away!"

Following Don's advice, I wrote out "1D10T."

"What does it spell?" Don said, straining to contain himself.

"Ahh... ummm... IDIOT?"

The two of them laughed so hard I thought they were going to pass out. Don fell off his chair and rolled around on the floor; tears streamed down Alan's face and his whole body shook convulsively. I stared at them incredulously until my anger finally yielded to the grudging realization that, as usual, they were right. It suddenly dawned on me that if I kept working at this pace, I was going to turn into another one of those strident, cheerless, cynical conservationists I had vowed never to become.

Fortunately, while my own and several other scientists' cooperative research programs on the refuge ebbed and flowed, Hakalau's restoration program had kept chugging along. After being away for several years, I got a chance to go back up to the refuge and take a guided tour with Jack Jeffrey in December 2005. As we slowly wended our way up the Saddle Road out of Hilo, Jack began with an oral update:

> We're still planting koa—about twenty thousand seedlings a year now—but we stopped growing them in our greenhouse because we've shifted a lot of our efforts towards growing and planting out more of the other species, and we just don't have enough space to do both. We'll probably grow and plant out fifteen thousand non-koa seedlings this year, including five hundred endangered plants. So now we just harvest the koa seeds and give them to a local nursery. This year we contracted them to grow fifteen thousand koa seedlings for us, but somehow they ended up with thirty thousand! So we bought another ten thousand, and Baron almost hit the roof. He said, "Wait a minute, you promised we were going to slow down this year!"

Jack told me that they were no longer limited by the number of volunteers they could recruit, and in fact they were now booking groups more than a year in advance and had to turn some people away. The two most important factors limiting their restoration program were simply money and time:

For example, we desperately need more staff, and we're going to have to start cutting back on our plantings due to our budget. The koa seedlings used to be around a quarter apiece, but now they're getting close to a dollar. But what we really need is a volunteer coordinator to organize them and spend all those hours answering e-mails and phone calls and solving all the logistical problems. We've had several 'volunteer' volunteer coordinators, but they don't last because it's really a hard, full-time job, and eventually their enthusiasm runs out.

We commiserated on the frustrations of working for the government and dealing with the endless politics. Jack was particularly upset by the amount of federal money being spent on what he (and many others) considered frivolous, politically motivated projects while the restoration work at Hakalau and throughout Hawai'i in general limped along on shoestring budgets. He also confessed that he and several of his colleagues had been working so hard for so long that they were all starting to get burned out (apparently no one had told them about the 1D10T code).

But on a happier note, their program had finally begun to receive some long overdue recognition, and Jack himself had recently won several prestigious honors, including the National Wildlife Refuge System's "Employee of the Year" award and the Sierra Club's "Ansel Adams Award" for making "superlative use of still photography to further a conservation cause." (In addition to their frequent appearances in books and magazines, Jack's spectacular photographs often show up in both technical and lay-audience oral presentations. As I discovered when I first began giving my own talks about Hawai'i, Jack readily gives his pictures away to anyone using them for research, conservation, or education purposes.) Nevertheless, despite such honors, the refuge was still as pathetically short staffed and underfunded as ever.

"However," Jack said, brightening, "there is a 'Friends of Hakalau Forest National Wildlife Refuge' organization in the process of getting chartered, mostly by the people who have volunteered in our tree-planting programs over the years, and maybe they'll be able to funnel some independent money to us that we'll be able to use to hire a volunteer coordinator. And the political lobbying and hunter pressure to manage for more ungulates is finally starting to ease up a bit."

When we turned off the Saddle and onto the dirt road leading to the refuge, I was relieved to see that it had finally been partially repaired and wasn't quite as washed out as it used to be. But my mood darkened again when I saw the gorse.

"Yeah," Jack sighed when he saw my expression, "it's gotten a lot worse since you were here. There's probably over six thousand acres of it now, with forty to fifty thousand stems per acre." We drove along through the miles of gorse in gloomy silence for awhile until Jack resumed his story.

> We've found that just about anything we plant under the koa corridors will survive; the big problem is the grasses. We had hoped the koa would eventually shade them out, but even in our oldest plantings with the most closed canopies, the kikuyu grass is still there. Not four feet high anymore, like it was, but brighter green than outside the koa due to all their nitrogen fixation and still dense enough to choke out everything else. We've been experimenting with spraying a grass-specific herbicide in the corridors and then throwing out some seeds and seedlings, but so far it doesn't look very promising.

Jack told me that a few years earlier, a fire had burned away most of the grass in one of the refuge's transition forests that contained some older but sparsely distributed koa and ʻōhiʻa trees. At first they were excited to find a lot of native seedlings coming up in the absence of the grass:

> "Hey," we said, "pretty neat, there's still a native seed bank there!" I think this was because there are still ʻōmaʻo [*Myadestes obscurus*, the Hawaiʻi thrush] flying through there and dispersing the native berry plants. But then we went back a few months later and it was just solid kikuyu again, and as far as we could tell not a single native seedling had survived. But now I'm much more hopeful. If you look around the

base of some of the larger trees, you see that the leaf and bark litter is starting to accumulate, and then some of the native ferns come in and establish on that litter, and then eventually you start seeing native plants coming up within those ferns. In areas where there's no kikuyu, these little *kīpuka* [islands] of native species are moving towards each other at a rate of about three or four inches per year. It's also happening in some areas where there is kikuyu, although much more slowly. So maybe someday the native plants will eventually shade out the grasses. And in some of the more intact, old-growth forests without the kikuyu, we're seeing an explosion of regeneration by some native species that had previously been wiped out by the cattle and pigs. For example, just this year the *ōlapa* is starting to fruit en masse. I learned over many years of photography that in order to recognize what's going on and appear in large numbers, the birds seem to need a critical mass of flowers and fruits. Of course, in the old days there would have been things like *ʻōʻū* [a once common honeycreeper that is now presumed extinct] and *ʻalalā* [*Corvus hawaiiensis*, the Hawaiian crow that is now extinct in the wild], but now *ʻōmaʻo* are the only native bird left that can disperse the native seeds. If they disappear . . . [Jack shuddered at this thought]. Anyway, there's no reason for the *ʻōmaʻo* to go into our corridors yet because the koa themselves don't provide much for them to eat. But hopefully, if the understory stuff we've planted in there takes off and starts fruiting, they'll come.

Jack explained that after seeing little natural native plant establishment within their koa corridors, they realized that at least for the time being they were going to have to jump-start this understory regeneration themselves. They hoped this would eventually create both a seed bank for the plants and a food bank for the birds. Ideally, as the koa matured and created more shade and organic litter, these plants would create the little nongrass *kīpuka* they observed elsewhere on the refuge, and the birds would disperse the seeds and accelerate the native plant expansion.

"Things progressed more slowly at first than we would have liked," Jack admitted. He proceeded to explain:

Remember all those 'ōhi'a we planted in the corridors back in the mid-1990s? We're lucky if we can find one now that's even three feet tall. Even the other faster-growing, more shrubby stuff has generally grown very slowly. Is it from root competition? Reduced light from all that grass in the understory and koa canopy overhead? We don't know.

Up until a year or two ago, I was pretty pessimistic about our chances of seeing the native plants colonizing and establishing out there on their own, but now I'm much more optimistic. And I also thought the non-koa plantings were still too small to have much of an effect on the birds. But just the other day I was heading down to the old-growth forest to do a bird survey. I was shivering out there in the pasture next to one of the corridors, waiting for the dawn to come, when I heard two 'elepaio [*Chasiempis sandwichensis,* a small native flycatcher] less than ten feet away! Now they didn't come all that way up there that morning—I was a half mile from the forest—so they must have roosted in the corridors. And then not so long ago I was driving around the pastures and heard a juvenile 'aki [see Plate 7, *Hemignathus munroi,* the endangered 'akiapōlā'au honeycreeper with the famous "Swiss Army knife" bill]! Apparently some 'aki family groups are regularly using the corridors for feeding. And now you can sit on the porch at the refuge cabin and see 'amakihi, 'i'iwi, and 'apapane [*Hemignathus virens* (Plate 8), *Vestiaria coccinea,* and *Himatione sanguinea,* respectively, three of the more common honeycreepers].

I never would have predicted that the birds would respond to our restoration efforts that quickly, but to me this proves that we are doing something right. Of course, we don't know what is going to happen as the corridors age. For example, 'ākepa [*Loxops coccineus,* an endangered honeycreeper that nests exclusively in tree cavities] is an old-growth forest bird; we still haven't seen any in the corridors.

But maybe if we're lucky, we'll start seeing some in the transition zones between the old-growth forests and the corridors. Same thing for the creepers [yet another endangered honeycreeper]—they need rough bark to find insects, but the koa in the corridors are still too young and have only smooth bark.

When we finally emerged out of the gorse and reached the entrance to the refuge, I hopped out and opened the gate. As Jack drove through, I examined a big muddy pig wallow that extended right up to the refuge fence line as if to say, "Sure, you may have excluded us for the time being, but we're still here, watching and waiting." However, immediately after I closed the gate, a pair of *nēnē* waddled over and serenaded me with their eerie, muted calls that sound like a cross between a sick cow and a distant foghorn. As we drove farther into the refuge, I saw *nēnē* everywhere. Jack told me that thanks to their recent concerted predator control efforts, the refuge flock, which had been founded in 1996 with eight geese, was now well over a hundred.

When we reached the greenhouse, I saw plants stuffed into every conceivable nook and cranny inside the building and many more outside scattered on the ground all around it. They had also converted what had been the outdoor pens of an adjacent pig-hunting dog kennel into a series of makeshift shade cloth structures to house some of their larger potted plants. (They got rid of the dogs after they started going after the *nēnē*.) Because the plants were so vigorous, it was easy to forget that these were some of the rarest and most endangered species on earth. We walked over and went inside to take a closer look. As usual, Jack had great stories to tell: This *Cyrtandra tintinnabula* (a genus from the African violet family with fifty-eight species found only in Hawai'i; only the genus *Cyanea*, with seventy-three unique species, has more) was presumed extinct until he stumbled across some while running a vegetation transect; that beautiful mint, *Phyllostegia brevidens*, was last seen in the 1800s until it was rediscovered on the refuge in 1990. Although this species is now extinct in the wild, to date they had planted out about two hundred of them.

When I asked if they were having any problems with inbreeding, Jack just shrugged. "Probably, but there's nothing we can do about it."

I walked over to get a better look at some of their strange yet exquisite lobelias, the "peculiar pride of our flora," as the pioneering German botanist W. F. Hillebrand had called them in 1888. Jack pointed to a *Clermontia pyrularia*, a lovely, Dr. Seuss-like tree about ten feet tall. He told me that one of the refuge staff members accidentally found a wimpy-looking individual one day while slogging through a dense patch of alien banana *poka* vine. Over the next few years, Jack laboriously searched in vain for more specimens, until one day a colleague found a bunch of much larger ones in that same banana *poka* patch. "I never thought to look up that high," he laughed. "Who knew they could grow so big and tall in there?"

Jack showed me yet another endangered lobelia, *Cyanea shipmanii*. Like all Hawaiian lobelias, the leaves on this small, bonsai-like tree were borne as tufts at the ends of its stems and branches, and its curved and tubular flowers strongly suggested a close coevolutionary history with Hawaiʻi's similarly curve-beaked, nectar-feeding forest birds. Unlike almost all of Hawaiʻi's other unique plants, however, this one had sharp spines and other physical defense structures. Interestingly, as is the case with several other species within this genus, these armaments were much more abundant and better developed on the lower, younger parts of the plant and less pronounced or absent completely on the upper, older sections. Researchers have found that this transition from more to less physical protection generally occurs in *Cyanea* species around three feet off the ground. Since the maximum vertical reach of Hawaiʻi's prehistorically extinct, giant herbivorous geese and their gooselike relatives is projected to have been around four feet, some have intriguingly argued that the spatial distribution of these physical armaments evolved as a defense against what may have once been intense avian herbivore pressure.

While Jack and I were talking, a bunch of *nēnē* had wandered over, and they suddenly began honking all around us. As I watched them, a vivid vision of a vast, primordial flock of flightless geese flashed through my mind, and a powerful wave of chicken skin washed over my body.

"Yeah," Jack said, looking at the *nēnē* fondly, "they love coming around here. Sometimes some of the honeycreepers fly over, too, and try to get at the flowers through the shade cloth."

We got back in the van, rolled down the windows to enjoy the beautiful afternoon and listened to the chorus of native birdsongs, and headed down to a thirteen-year-old koa corridor Jack wanted to show me. When we reached our destination, I walked over and was stunned—it was a real forest! I gazed at the sprawling koa canopy thirty feet above me, then studied the impressive diversity of transplanted native species scattered beneath them. The restoration was so impressive that when I walked over to examine the adjacent koa corridor, it took me a minute to figure out where the one I was in stopped and the next one began.

"As you can see," Jack said when he caught up with me, "they're starting to fill in with koa seedlings and root suckers coming off the established trees. In fact, the day may soon come when we stop planting koa altogether and just do the understory and endangered species. And the kicker is the ferns are starting to come in on their own! We've seen them popping up here and there, but mostly they seem to like it beneath the koa, especially in the extra shade around the 'ōhi'a we've planted beneath them. And look at that," he said, pointing to several clumps of a *native* bunchgrass growing in an area where the alien grasses weren't so thick, "*Deschampsia nubigena.*"

We walked back into the heart of the first corridor and just watched and listened to the birds for a few minutes. "There's the mother of the rain forest," Jack finally said, pointing to a thick patch of *uluhe* fern that appeared to be holding its own in the thick surrounding kikuyu grass. "Eventually the other natives are going to come up through those ferns just like they do in the old-growth forests. Even a year ago I would have predicted that there's no way we would ever get this much volunteer native establishment in here."

We walked out of the corridor toward the van, then turned around for one last look. Jack concluded, looking at the trees,

> You know, this whole thing has worked much better than
> I ever dreamed—I never thought I would live to see and
> hear all this. Of course, as you've seen, lots of these koa have
> multiple trunks and branches that spread out like a palm
> tree. They don't grow up straight like the canopy emergents
> in the old-growth forests. We've been criticized for this, but
> there just ain't nothing we can do about it. Same thing with

ʻōhiʻa—we're growing multipodial rather than monopodial trees. We know that we need old, monopodial *ʻōhiʻa* that are at least two feet in diameter to develop the kind of rot spots that are suitable for *ʻākepa* nesting. But at the growth rates we see out here, that'll take about six hundred years! My feeling is this is not the final stage; this is not the final forest. In the future, when we've got the canopy, then the other trees will start filling in, and then we can get the normal, natural form.

As we drove back up the mountain, I asked Jack whether he was optimistic or pessimistic about the future of Hawaiʻi's native biodiversity in general. He laughed,

I'm pessimistically optimistic! In the beginning, I was very pessimistic because I was going to areas where the birds were there in the not-too-distant past, and now they were effectively gone. I saw the last *olomaʻo* [*Myadestes lanaiensis,* the Molokaʻi thrush]; I heard the last *ʻōʻō* [*Moho nobilis,* a honeyeater found only on Kauaʻi]. The last *ʻōʻū?* The last *kāmaʻo* [*Myadestes myadestinu,* a thrush found only on Kauaʻi]? Probably. Seeing the birds dropping out right in front of me made me very much a pessimist. But then I got into restoration, and I started to see we can do things that make a big difference, and we can turn this whole thing around. Will we be able to save everything that we've got left today? Probably not.

Nevertheless, my hope is that my great-great-great-great grandkids will be able to come here and see the birds I'm seeing now. I know that we can't turn everything into a refuge like Hakalau. Eventually we are going to end up with little islands of protected areas within these islands. I hope they aren't too small, and I hope we will have some sea-to-mountaintop *ahupuaʻa* [a Hawaiian land division system somewhat similar to our modern concept of ecological watersheds] that are managed for native species on all the Hawaiian Islands. Whether or not that happens will ultimately depend on the public—we've seen in the past just how quickly the politics and the pendulum can change.

But the success we're having here at Hakalau is very encouraging. And of course, as you saw and heard, the best part is that the birds are coming back. It takes a lot of time, and education, and hard work, but I guess that now I can say that yes, by golly, if you plant it, they really will come.

Part 2 **Restoration Roundup**

6   KILL AND RESTORE
Hawai'i Volcanoes National Park

Hawai'i Volcanoes National Park (HVNP) literally became my backyard after I moved to Volcano Village in 1997. I came to know and love the park's spectacular 323,431 acres (an area comprising over 12 percent of the Big Island and nearly 8 percent of the entire state) through many subsequent hiking, running, and camping adventures. I also participated in professional activities there, such as helping other scientists with their research, conducting VIP tours, and contributing to resource management advisory sessions. Over time, I developed close relationships with some of the park's employees and came to increasingly appreciate their accomplishments and dedication toward preserving and restoring native species and ecosystems.

For many years, however, all I knew about the park's history was what I read on one of their tourist displays: "Hawai'i Volcanoes National Park was established in 1916 primarily for its accessible volcanic scenery and value for geologic study." Later, when I became interested in learning more about how this park became such a showcase of effective ecological restoration, just about everyone I asked told me to go talk to Don Reeser. One award-winning scientist and conservationist put it this way:

> Don Reeser is one of my heroes. He single-handedly turned around the entire management paradigm at Hawai'i Volcanoes National Park by getting rid of the goats. The state was totally against it; the Park Service administrators were against it; an early head of the Forest Service in Hawai'i argued the native species were hopeless and advised the state to give up on them. But Reeser didn't care—he didn't respect their science or their judgment—he just went out there and did it.

When I finally caught up with Mr. Reeser, who had since moved to Maui and worked as the superintendent of that island's Haleakalā National Park from 1988 until retiring in 2005, he seemed to remember his early years at HVNP as if they were yesterday. He began by modestly telling me that when he transferred to the park in 1968, there was already a long tradition of goat control there. In fact, he pointed out that their staff had attempted to control its feral goats since its creation and had killed or removed over seventy thousand of them by the time he arrived. Yet although such data were often used to demonstrate the effectiveness of their goat control program, Reeser and his on-the-ground colleagues knew this claim was patently false. "The truth," he told me matter-of-factly, "was that the goats were still extremely abundant throughout a large section of the park, and anyone with half a brain could see that they were still wreaking havoc."

He showed me an article he had written that summed up his assessment of the situation at that time. Mincing no words, he stated,

> In actuality, this "success" did virtually nothing to protect native flora and fauna. In 1970 an aerial census revealed there were over fourteen thousand goats in the park, probably as many as there had ever been. Drives and hunts were only effective in keeping the goat population young, healthy, and vigorously reproductive. My predecessor . . . and others before him had prescribed boundary and internal fences

as the only solution to the goat problem, but their well-conceived plans gathered dust in the files.

Much like the early struggles of the Hakalau Forest Refuge staff to garner the necessary administrative, economic, and logistical support to fence and drive out their feral pigs and cattle, Don told me how the ongoing pressure from the politically powerful hunting lobby forced the Park Service to rely on volunteer citizen control programs. But as the ecological conditions in the park continued to deteriorate, increasing pressure from the conservation community finally led to a request from the assistant secretary of the interior for a report analyzing the goat situation at HVNP. Don and his Park Service colleague James Baker eagerly worked on this report while forging ahead with fencing projects designed to demonstrate the ecological benefits of protecting native plants from goat browsing.

Their report, "Goat Management Problems in Hawaii Volcanoes National Park: A History, Analysis, and Management Plan," published by the Park Service in 1972, begins with this quote from the famous Leopold report entitled "Wildlife Management in the National Parks" (1963): "A visitor who climbs a volcano in Hawaii ought to see mamane [*Sophora chrysophylla*] trees and silverswords, not goats." Given today's ongoing political battles about the ecological effects of ungulates and the hunting and ranching lobbies' frequent claims that they have not received sufficient scientific study and remain poorly understood, I was struck by how well Reeser and Baker understood this problem four decades ago. For example, after presenting several detailed case histories that demonstrated how destructive feral goats have been in the world in general and on islands in particular, their report goes on to detail the following scenario:

> The many instances of goat/island introductions and subsequent vegetative destruction lead to the conclusion that when goats are present in large, uncontrollable numbers their effect on native flora is considerable and biologically disastrous. Several things may happen . . . when goats or other hoofed animals are introduced into an island ecosystem. First, the herbivorous ungulates are introduced

into habitats that never before knew pressures from such animals. The vegetation had usually lost or never evolved protective mechanisms such as spines, thorns, chemical repellents or irritants, and unpalability. Also, the vegetation may have no defense against or adaptability to recover from excessive trampling. The newly introduced herbivores, without any natural predators except man, may increase rapidly.

They then focus on the effects of feral goats in the western Pacific and Hawai'i and quote several earlier authors who also clearly realized how destructive these and other ungulates had been to the islands' native species and ecosystems. They conclude their introduction with another remarkably prescient passage that illustrates their awareness of the other major threats to the long-term survival of the park's native biodiversity:

> Goats in Hawaii have, therefore, been long recognized at the state and federal levels as significant factors in vegetation and species destruction. In less than 200 years, goats have chewed their way from the seashore to the tops of island peaks and back down again, by which time they numbered in the tens of thousands, had eaten some species into extinction, and threatened the existence of many more. Moreover, the resultant forest destruction has been one of the major causes for declines in populations of native, endemic, nectar-feeding birds. . . . Goats, however, are not the only resource problem within the park. To lesser degrees, feral pigs, nonnative rats, introduced mongooses, and exotic birds are most assuredly creating problems, however subtle, that affect native species and habitat in many ways not yet determined. . . . Probably second only to goats, nonnative plant invaders create the park's most pressing management problems, and exotic plants may well remain problems long after wildlife dilemmas have been resolved. . . . It is ironic that while the goat has been recognized in the past to be the most troublesome and most pressing of the park's many problems, it may now be the most easily solved.

Later in the report, Reeser and Baker argue that the park could and should eradicate their goats through a combination of fenced management units and intensive and coordinated professional hunts and goat drives (i.e., use the same "divide and conquer" methodology that the Hakalau Forest staff employed so successfully). Yet much like the Hakalau experience some years later, Don and his colleagues' ungulate management plans were thwarted by political pressures from the hunting community, which successfully lobbied the Park Service to employ "citizen participation goat control programs" that effectively maintained abundant goat populations and sustained-yield recreational hunting within the park.

Don told me that many Park Service staff and research scientists in those days believed that the goat and cattle damage was so severe that it was probably irreversible, and that these ungulates should in fact be retained as "a necessary evil" to control the ever-worsening alien plant invasions and to reduce the risk of wildfires by browsing down the exotic grasses that dominated many of the drier sections of the park. While he knew that the goats and other ungulates ate large quantities of weeds, he was skeptical that their foraging was really slowing down the spread of alien plant species and benefiting the native species.

To investigate these kinds of questions, Don and his colleagues initiated a long-term study in 1969 that included a series of exclosures to monitor the fate of native plants in the absence of goats and other exotic animals. To the great surprise of many at that time, within a few years their exclosure data showed that many native plant populations could recover and at least coexist with alien vegetation in the absence of ungulates. In their 1972 report, they summarize the results of those first goat exclosures and other complementary studies within the park by drawing six "startling conclusions":

- Goats are selective and choose native species over nonnative species whenever available;
- Goats help to perpetuate the growth and spread of less palatable exotic species;
- Goats can, in time, deteriorate a native forest into a grassland-savannah;

- Goats are the likely cause of extinction, or near extinction, of several species of soft-barked native trees and shrubs;
- Goats denude areas of their vegetative cover to a point that causes shallow topsoils to erode, exposing rocky substrates; and
- If goats can be removed from the scene, native plants will reappear, flourish, and in some cases out-compete invading exotics.

Their most dramatic results came from an exclosure located within a dry, desolate region of the park that also contained the highest goat densities. Prior to fencing, there was not a single native plant within this exclosure, and its two dominant alien grass species were "so overgrazed that the ground cover resembled a closely mowed lawn." Yet after only two years of goat exclusion, they found that several native species had reappeared and occupied more than half of the ground cover. But what really got everyone's attention was the appearance of what at the time was considered a new legume species—*Canavalia kauensis*. When Reeser and Baker subsequently found more dormant but viable *Canavalia* seeds lying around in the area surrounding this exclosure, they hypothesized that the absence of goat browsing had allowed the dense growth of both native and alien grasses inside the fence, which in turn protected the ground from the desiccating winds and enabled the soil to retain enough moisture for these *Canavalia* seeds to germinate, establish, and eventually even crowd out some of the exotic species.

The miraculous emergence of this phoenixlike species (because later analyses concluded that these plants' unique morphological features did not warrant formal taxonomic status, they were ultimately reclassified as a nondistinct variant of a more common species), along with photos showing a relatively lush grassland inside the fence and a rocky, desolate, goat-infested landscape outside of it, helped fuel an international barrage of protest against the Park Service's resource management policies. Consequently, the political tide finally began to turn: New administrative leadership was brought into the park; funding for ungulate control was aggressively pursued and obtained;

Goat exclosure at Hawai'i Volcanoes National Park in the 1970s. *Don Reeser*

and comprehensive ecosystem protection and restoration plans were developed and implemented.

In 1974, the first official Resources Management Division was created at Hawai'i Volcanoes National Park. (Similar professional divisions were eventually created throughout the US National Park System.) This group, originally comprised of locally hired laborers who were familiar with Hawai'i's physical and cultural terrain, would often camp out in the field all week to build fences and control alien plants and animals. Reeser told me that he thought the employment of these local folks and their heroic dedication to their mission were the key factors that ultimately defused the opposition to this division within and outside of the Park Service.

Reflecting back on this work years later, Don wrote,

> Taking out the first perhaps 80 percent of the goats or so was easy. The last few goats in a unit required super

determination. We had to continually remind ourselves that removing only a few goats during a hard day's effort was not failure but a sign of success. Frequently we would celebrate the removal of the last goat prematurely. The last few survivors in a unit were exceptionally wary and extraordinarily elusive. . . . By our dogged persistence, unit by unit, the goat population declined.

To flush out and kill the last few goats in the most remote and rugged areas, they utilized a combination of discreetly executed helicopter hunts and radio-collared "Judas Goats." Haleakalā National Park eventually applied similar strategies against their goat populations, and by 1990 goats were effectively eradicated from both of Hawai'i's national parks.

After the dramatic successes of their goat control programs in the 1970s, the park shifted its efforts in the 1980s and 1990s toward managing its feral pig and alien plant populations. Unlike the goats, which they eradicated throughout the entire park, the resource managers knew it would be virtually impossible to control the pigs and weeds on such a vast scale. They thus decided to focus on the relatively intact regions that appeared to have the greatest potential for ecological recovery.

For the pigs, such areas largely turned out to be the sections that still contained significant coverage of native canopy trees. By the end of the century, the park had eradicated every last pig within ten units encompassing a combined area of approximately twenty-five thousand acres. As has been the case elsewhere throughout the Hawaiian Islands, the ecological recovery of these relatively intact forests following the removal of their pigs was generally rapid and substantial. The fences have also been effective at keeping the pigs in the surrounding areas out; on average, they have detected only one "pig ingress incident" every two to three years within each of these control units (these invading pigs are then quickly killed by standard dog-hunting, snaring, and trapping techniques).

The park is now evaluating the feasibility of additional control programs within some shrubby areas and grasslands that contain low and potentially insignificant pig densities, forests with dominant alien plant understories, and relatively intact native forests that are difficult to fence due to volcanic activity or difficult to hunt due to extensive cracks in the underlying lava substrate.

Although their goat, pig, and other ungulate eradication efforts were all extremely difficult, laborious, and expensive, they were a relative breeze compared to the challenge of controlling the park's alien plant populations. As succinctly stated in an internal Hawai'i Volcanoes Resource Management document, "The spread and control of alien vegetation present a complex and only partially solvable problem." This report goes on to explain that this sobering reality is due to many factors, including the following:

1. The park is currently infested with approximately 400 different alien plant species, and the distribution and abundance of at least 24 of these weeds is too great for meaningful chemical or mechanical control.
2. Several of these species already have or are capable of invading and displacing even relatively intact stands of native vegetation and altering basic ecosystem properties such as fire and nutrient cycling processes.
3. Some of these weeds are at present rapidly expanding their range and density within one or more sections of the park.
4. "Eradication" of many other less prolific, seemingly manageable alien plant species may also be difficult or impossible due to these species extensive, persistent soil seed banks within the park and/or ongoing reintroductions of new seeds into the park.
5. The park's overall size and frequently rugged, inaccessible terrain make detecting and controlling new alien plant invasions difficult.
6. Some effective mechanical and cultural weed control techniques cannot be used because they would not be appropriate within a national park and they could damage the park's existing native species as well as its culturally important archaeological sites and attractive physical features.

Nevertheless, despite these and many other challenges, the park's ongoing war against the weeds is neither hopeless nor futile. As an alien plant control field supervisor told me,

> Sure, there are times when it all seems overwhelming and pointless. But because we've mucked it up, we have a moral obligation to try and fix things. So we analyze the situation, formulate our strategy, make some tough decisions, present and defend our logic, and get to work—you just have to start somewhere and give it your best shot. Along the way we've learned to accept the fact that we may be wrong, so we monitor ourselves as objectively as possible and remain ready to change and adapt as necessary.

Because they can't employ an "eradicate and fence" operation for weeds, the park's alien plant strategy is necessarily more nuanced and multifaceted. As is the case with most conservation programs throughout the state, one of the overarching goals is to minimize the kinds of ecosystem disturbances that facilitate the establishment and spread of invasive alien plants. In the park, the two most important agents of this type of disturbance are ungulates and fire. Consequently, ungulate control remains the top resource management priority and almost always precedes any weed management activities. They have developed an extensive fire prevention program and employ an aggressive fire suppression policy. Ironically, however, the red-hot lava flows that helped create the park and remain its biggest tourist attraction are also a major and probably intractable source of the kinds of fires and physical disturbances that can destroy native ecosystems and facilitate the invasion of fire-promoting exotic grasses (see Plate 9).

In addition to minimizing and mitigating the disturbances they can control, the alien plant crews map and monitor the most important weeds, deploy manual, mechanical, and chemical control techniques against selected weeds in strategic areas, participate in interagency partnerships to control potentially invasive new species, strive to improve their ecological understanding of key alien plants, and work with other agencies and individuals to develop new and improved weed control tools. For example, several of the scientists

Park Service staff install temporary fencing around native plants.
National Park Service

and technicians in my Forest Service unit in Hilo work at a quarantine facility for entomological biocontrol research within the park under a memorandum of understanding signed by these two federal agencies, the University of Hawai'i, the Hawai'i State Department of Land and Natural Resources, and the State Department of Agriculture.

The park has also employed some more aggressive alien species control and native species restoration activities within a series of so-called Special Ecological Areas. These SEAs, which are intensive management units in which the park employs techniques that are not feasible on a parkwide basis, grew out of a desire to preserve and restore some of their most ecologically valuable sites while they were still relatively intact and manageable. This program began in 1985 with the establishment of six SEAs encompassing over nine thousand acres, then expanded over the next five years to include twelve SEAs comprising almost thirty thousand acres and multiple sites within four of the park's six unique ecological zones.

The rapid expansion of their SEA network was made possible by the progressively declining workloads following the initially intensive alien species control efforts. For example, the park's crews needed approximately 1,019 worker-days to control and monitor the most important weeds within three representative SEAs during the first year, 486 worker-days during the second year, and only 210 during the sixth year. The abundance of many of the most noxious weeds within these SEAs also dropped by one or sometimes even two orders of magnitude during this time period. By the time I first saw them in the late 1990s, some of these SEAs had been transformed from impenetrable weed jungles to nearly weed-free forests.

The park is also committed to the preservation and restoration of its four charismatic flagship species: the *nēnē* goose, the hawksbill sea turtle, the dark-rumped petrel, and the Mauna Loa silversword. In addition to the ecological and ethical reasons for saving these endangered species, several staff members told me that their continuing presence generates positive public relations because most of the public can relate to and empathize with these beautiful yet vulnerable organisms. For many people, seeing these species or just knowing they are there also provides an emotionally powerful demonstration of the value of nature preserves (they never failed to inspire me when

I encountered them!) and creates opportunities for them to educate their visitors about more controversial issues, such as the need to kill alien species and aggressively manage "natural" ecosystems. While each of these flagship species has its own unique challenges, their collective recovery largely involves yet more plain old alien species control and native plant restoration programs. For instance, the park's goal of increasing its *nēnē* population from two hundred to five hundred birds heavily depends on intensive predator control programs (primarily pigs, mongooses, and feral cats) within the *nēnē* breeding territories. Because recent studies have suggested that inadequate nutrition may be a major limiting factor, they are also attempting to restore nutritious native forage plants across strategic sections of the park's lowlands.

The park's population of pelagic dark-rumped petrels (*Pterodroma phaeopygia sandwichensis*) is largely confined to the higher elevations of Mauna Loa because the lower-elevation petrels, along with the other seabirds, were probably extirpated by feral cat, mongoose, and rat predation. However, a few wide-ranging feral cats now appear to be attacking these once predator-free, high-elevation petrel nests and significantly limiting their population size. Current research is thus attempting to find more effective ways to control these elusive cats and improve our understanding of the petrel's distribution and population demography. The park is also studying the feasibility of reintroducing several locally extirpated endangered native forest birds at lower-elevation sites on Mauna Loa.

The hawksbill turtle (*Eretmochelys imbricata*) is one of the most endangered marine turtles in the world. As far as we know, it regularly nests only in the Hawaiian Islands, and forty-seven of this species' fifty-three observed nesting sites are on the Big Island. Over the last decade, members of the park's staff have found nine nesting beaches on the island, three of which occur within their boundaries. Extensive research has since revealed that the major factors limiting their nesting and hatchling success appear to be mongoose predation and hatchling strandings due to altered beach conditions, disruptive human activities such as fishing, beach vehicles, and artificial lights, and the establishment of alien plants. To help ensure the continued survival of this species, throughout their seven-month breeding season

Hawksbill turtle hatchling (*Eretmochelys imbricata*).
National Park Service

park staff continually search for and protect the turtles' nests, control their predators, and assist their stranded hatchlings. Because the majority of their nests occur outside the park, they have also partnered with other groups to help educate the public and foster their support for turtle monitoring programs and coastal conservation.

The long-term survival of the park's only non-animal flagship species, the ethereal Mauna Loa silversword (*Argyroxiphium kauense*), will also probably depend on the success of its interagency partnerships. This is because this species is presently restricted to three small, relictual populations outside their boundaries on the upper slopes of Mauna Loa. Portions of these populations are protected by fenced exclosures, but mouflon sheep and feral pigs and goats are abundant throughout this area. Because the park has large expanses of suitable silversword habitat on Mauna Loa that are already fenced and managed for ungulate control, it has begun a massive cooperative effort to collect and propagate thousands of silversword seeds from these remnant populations and outplant the resulting seedlings into smaller fenced exclosures within their existing network of larger fenced areas on Mauna Loa. These smaller exclosures are necessary because mouflon sheep have not been completely eradicated in this region, and just one of these animals can quickly destroy an entire population of small outplanted silverswords.

Much like the native wildlife at Hakalau, the plight of these four flagship species—and the native species and ecosystems inside the park in general—can also be viewed as a microcosm of the current and future status of Hawai'i's biodiversity as a whole. On the one hand, there are good reasons to be cautiously optimistic that all of these flagship species will survive and possibly even thrive into the foreseeable future. Indeed, many of my colleagues and I have had the pleasure of witnessing just how quickly some of the park's native species and ecosystems can rebound when given a fighting chance. Some believe that at least a few of these areas have already been effectively restored, because the seemingly unstoppable tide of alien species has been pushed back and key native species have reclaimed the newly opened spaces via both natural recruitment and deliberate reintroductions. The park also seems well positioned for even more ecological success in the future: They now have substantially more hard-won knowledge about how to control their alien pests and restore their native species and ecosystems; more experienced and dedicated work crews; and a more fully developed physical and intellectual infrastructure, such as over one hundred miles of ungulate-proof fences and detailed GIS distribution maps for key native and alien species.

Of course, on the dreaded other hand, no one would dispute the fact that substantial portions of the park are getting worse, and some well-informed people fear their present ecological victories may be short lived. Many of their native species and ecosystems have not yet

Fence protecting Mauna Loa silverswords.
National Park Service

recovered despite the successes of their ungulate and weed-control programs, and the underlying factors preventing their recovery may be far more difficult to identify and address. For example, the activities of the introduced small mammals and invertebrates (especially yellow jackets and ants) are almost certainly major limiting factors, yet our technology to control these kinds of pests remains expensive, labor intensive, applicable only within small areas, and often only partially effective at best. In other cases, it may be that the factors inhibiting ecological recovery are intractable either because they are beyond the scope of our ability to detect, understand, and address (some change in the soil microbial community, an undescribed and untreatable disease/parasitic interaction) or because they are caused by the absence of one or more formerly crucial native species that have since become functionally or actually extinct.

Finally, there is always the possibility of detrimental political and physical change in the future. Just maintaining, let alone expanding, the park's current alien species control and native species restoration programs requires a substantial and continual influx of money, administrative commitment, and on-the-ground dedication and expertise, and there are no guarantees that any of these commodities will remain sufficiently available in the long-term future. There is also the very real possibility of potentially devastating future physical changes such as a series of new, even worse alien species invasions, severe human-induced climate change, or a series of catastrophic volcanic eruptions that could destroy some or all of the best remaining native ecosystems.

Nevertheless, at least for the time being, Hawai'i Volcanoes National Park remains a world-class biological jewel and a model for how effective ecological restoration can be when the necessary political will, resources, expertise, and dedication all come together.

My last visit to the park left me feeling more optimistic about its ecological future than I ever had before. The sun was peeking through the mist when I drove through the entrance gates, and the 'ōhi'a trees were in full sparkling glory. On my drive over to the Resource Management Division, I saw that large tracts of forest on both sides of the road had been thoroughly cleared of their formerly suffocating blanket of alien ginger and that native rain forest understory species

Post-ginger recovery of native forest understory. *National Park Service*

were popping up everywhere. When I reached their Administrative Headquarters, I almost couldn't find a parking space because there were cars and trucks and various Park Service vehicles everywhere. One of the senior crew supervisors told me there were now over sixty people working on the various Resource Management crews, which was far more than he had ever seen in all his years in the park.

Later, I grabbed a machete and went out into the field with one of the alien plant species control crews. We formed a line, spaced ourselves about fifteen feet apart, and chopped our way into the jungle of ginger that lay at the edge of a remnant native rain forest (see Plate 10). The work was hard and treacherous because the ginger was so thick that we couldn't see the ground, so if you were careless or unlucky, it was easy to wind up with a twisted ankle, bad scrape, or worse. I also heard more than one gruesome story about how someone had swung his razor-sharp machete at something and hit his own arm or leg.

Despite these physical difficulties, I was immediately struck by the crew's high morale, enthusiasm, and work ethic. And despite their considerable diversity of ethnicities, backgrounds, and ages, they all seemed united by love of their work and dedication to their mission. After a while I stopped to rest my out-of-practice, aching arm, wipe the sweat out of my eyes, and gulp down some water. I looked back at the impressive swath of ginger we had just cleared, then turned around and watched as the crew methodically chopped their way deeper into the forest.

"Why shouldn't these guys be happy?" I asked myself. They were out in the fresh air, working in a magnificent place, earning decent wages, and fighting the good fight. And at least for the time being, they seemed to be winning.

7   THE *PŪʻOLĒʻOLĒ* BLOWS
Dry Forest Restoration at Auwahi, Maui

In the botanical literature, Auwahi refers to a centrally located, 5,400-acre subsection of the southwestern rift of the Haleakalā Volcano on East Maui at an elevation of three to five thousand feet. Auwahi has been translated as "smoky glow" and "milk of fire," and the region is honored in an early Hawaiian song that begins, "Hot is Auwahi/Glowing, the lava of Hanakaʻieʻie."

The famous botanist Joseph Rock identified the remnant dry forests at Auwahi and North Kona on the Big Island as the two botanically richest regions in the entire territory of Hawaiʻi. The tragic story of the subsequent devastation of the native flora of both these areas has been remarkably similar: cattle ranching; imported, fire-promoting invasive grasses (fountain grass, *Pennisetum setaceum,* on the Big Island, and kikuyu, *Pennisetum clandestinum,* on Maui); and an ever-expanding tide of other noxious alien plants and animals. As he did in North Kona, Rock reportedly wept when he first saw the deterioration of the Auwahi flora in 1939 after being away for twenty years.

In 2004, Lloyd Loope, then the chief scientist at Haleakalā National Park, told me that for a long time the conservation community had lost all hope of ever restoring Auwahi:

Back in 1967, the Nature Conservancy had made the first attempt at conservation in Auwahi by constructing an ungulate exclosure, but then the kikuyu grass took off in there and it was just impossible to accomplish anything, so they eventually let the cattle back in and gave up. But then the sugarcane aphid eventually got in to Auwahi and started killing some of the grass, and the Native Hawaiian Plant Society built a series of small exclosures to protect patches of some remnant dry forest trees from the cattle. When I started working in Auwahi in 1981, I needed to find somebody who really knew their plants. Everyone told me I should hire a fellow named Arthur Medeiros.

I first got to know Art when he flew over to the Big Island to look at a dry forest restoration program in North Kona that I became involved with shortly after landing in Hawai'i in 1996. I had been amazed by the depth and breadth of his knowledge—not only did he seem to know every plant and bug and lichen we encountered, but he was equally informed and passionate about Hawaiian history, culture, and ethnobotany.

Before flying back to Maui, Art had invited me to come see his Auwahi restoration project. I took him up on this offer in the spring of 1999 when I flew over from the Big Island with Lyman Perry, a botanist for the Big Island's branch of the state's Division of Forestry and Wildlife. Lyman was going to Maui in part to hand-deliver four *māhoe* (*Alectryon macrococcus* var. *auwahiensis*) seedlings. This small, reddish-brown, federally endangered tree in the soapberry family had been relatively common on East Maui up until the early 1900s. Art told us that since an Auwahi *māhoe* he had been monitoring within a small exclosure had just died, apparently due to a combination of damage from alien black twig borers, termites, and the ongoing El Niño drought, there were now only eleven wild individuals left.

Fortunately, however, there was a *māhoe* tree in Manuka State Park on the Big Island that someone had planted many years earlier from seed collected at Auwahi. One of Lyman's colleagues spotted some intact fruits on this tree and gave them to another colleague who managed to propagate four healthy seedlings. This was fortuitous because most of these fruits are normally .lost to rodent and insect predation. The Hawaiians apparently also enjoyed them; Joseph Rock wrote that the "bright scarlet fruit flesh is eaten by the natives, as well as the kernel of the seed, and [they] are not altogether unpleasant."

When Lyman unwrapped the carefully packaged seedlings and handed them over, I saw Art's eyes well up. He held the pots in his trembling hands and silently studied the plants as if they were his long-lost children. Finally, he looked up and asked if we would like to help transplant them into some larger pots he had prepared that contained a mix of native Auwahi soil. Both of us shook our heads and left him alone with his *māhoe*.

I got a chance to see Art's restoration project again when I flew back to Maui in 2005. As we bounced our way out to his Auwahi exclosures along 'Ulupalakua Ranch's rough jeep road, Art briefed me on how his work and his thinking had progressed in the six years since I had been there. He began by explaining the origin of his now extensive Auwahi volunteer program:

> A coworker and I were having lunch at Auwahi, taking a break from some hard physical work and admiring the beautiful ocean. He looked over at the heavily degraded forest we were trying to restore and said, "Wow! What if this works?" I thought about that for a minute, then said, "What if it works and no one cares?"
>
> So I started recruiting volunteers. No one came at first, then they started coming—all by word of mouth—and now there are so many that, sadly, we have to turn a lot of them away. Every time I bake them cookies, and every time I share a piece of my soul with them, because this restoration is in some ways spiritual stuff, and you should get something out of it as well as give, so I work them hard, and I tell them they're doing something that's going to change the world: "That plant that you are planting, it's going to live longer than you are, and it's going to have more babies than you

are; it probably will change the world more than you are—its DNA is not going to go extinct—you're starting a DNA lineage on this land!"

Although at that time Art worked for the US Geological Survey, he was one of the very few other government scientists I knew who empathized with my struggles to simultaneously do both real science and real conservation. However, he was far less schizophrenic about balancing these two goals than I was. He told me flatly,

> Look, I love science; it's an essential tool. But it must be disciplined in order to be responsible, so that it also serves the land and the people. I'm not that interested in having a plump resume or advancing through the system; I'm more interested in conservation and on-the-ground accomplishments, having a *hālau* ["group," or literally, "a branch from which many leaves grow"], connecting with kids, mentoring young people, building community—that's the stuff that matters; that's what changes the world.

Art stressed that his Auwahi restoration program would never have gotten off the ground without the strong and enduring support of the landowners (the Erdman family of 'Ulupalakua Ranch) and their deep desire to give back to the land and the community that supports them. Eventually, the combination of Art's drive and the ranch's support led to the creation of a multiagency cooperative effort to restore a small but botanically rich section of Auwahi, which began with the installation of a ten-acre exclosure. By the time of my first visit to Auwahi, they had controlled most of the invasive weeds within that exclosure, constructed a greenhouse, and gathered and propagated thousands of native seeds.

In some ways, Art's coalition resembled our North Kona Dryland Forest Working Group on the Big Island. First, it was comprised of a diverse array of groups that included 'Ulupalakua Ranch, the US Fish and Wildlife Service (which, as was the case on the Big Island, had provided critical initial funding, in part because Auwahi contained many on-the-brink federally endangered species), Haleakalā National

Park, the Maui County Water Department, the Native Hawaiian Plant Society, Hoʻolawa Farms, Mahealani Kaiaokamalie and Living Indigenous Forest Ecosystems, and the Evolution, Ecology, and Conservation Biology Program of the University of Hawaiʻi at Mānoa. Second, they were largely powered by hundreds of volunteers. And third, they were interested in and motivated by both the biological and cultural aspects of dry forest restoration.

Unlike our consensus-driven, leadership-by-committee structure, however, the Auwahi Restoration Design and Implementation Team consisted of two: scientist Arthur Medeiros and ranch owner Sumner Erdman. "The entire Erdman family wanted to learn more about their lands and do the right thing," Art recalled. "Over time we came to trust and understand each other. We knew we wanted to try something bold to jump-start the restoration of a Hawaiian ecosystem in which natural reproduction for many of the native trees had not occurred in living memory."

For Art, "doing the right thing" involves not just carefully navigating his way through the seemingly endless labyrinth of ecological and philosophical conundrums so many restorationists struggle with but also devoting a substantial amount of thought and energy toward the cultural and spiritual components of his work. Thus in the year 2000, when his group had accumulated about twenty-five hundred native seedlings in their greenhouse and were ready to perform their first major outplanting within their ten-acre Auwahi exclosure, Art took a break from his hectic schedule to invite a famous Maui *kumu* (teacher) to bless their plants and welcome them back to Auwahi.

That outplanting day began with a group of about thirty people gathered around the greenhouse. After a round of prayers and "spontaneous words of inspiration," the *kumu* mixed ʻawa in a heavy wooden bowl and then sprinkled the ʻawa water around the greenhouse and plants before the now blessed seedlings were loaded onto the trucks that would carry them on their rugged forty-five-minute journey back to Auwahi. (ʻAwa [*Piper methysticum*], a shrub in the pepper family, was brought by the Polynesians to Hawaiʻi because of its cultural importance. Its roots were used to make a mildly narcotic drink that was and still is important for numerous Hawaiian social and religious occasions.)

When the volunteers reconvened beneath a spreading *kauila* (a rare native tree in the buckthorn family that produces exceptionally dense, hard wood that the Hawaiians used to make *kapa* (cloth) beaters, vicious spears (*ihe*), and poles for construction inside the ten-acre exclosure, another Hawaiian man blew the *pū'olē'olē* (conch shell) once for each of the four cardinal directions. Art later wrote that its

> loud brave cry filled the emptied forest, echoing off its rocky ridges. I found myself wondering how long it had been since the *pū'olē'olē* had sounded at Auwahi. One hundred years? Two hundred? Three hundred? More? Maybe that was the reason the dryland forest at Auwahi had fallen on such hard times! . . . As I watched the thin milky 'awa being poured from the coconut cup into the planting hole, I felt I was watching the *ola* (life) being poured back into the land. I had always thought the *ola* was in the plants, but now I felt the *ola* was in the land itself, awaiting the arrival of the seedlings.

Art also reported that near the end of the *kumu*'s blessing, the sky, which had been clear all week (and would be again throughout the week following this ceremony), began to darken, the ridge around the *kauila* tree became enshrouded in mist, and a steady light rain began to fall. They nearly canceled the scheduled helicopter operation to sling the seedlings in due to the rain and poor visibility, but then the pilot called and said that he felt comfortable and "somehow could see the things he needed to see."

"I'm not some New Age crystal-gazer," Art once told me after recounting a series of metaphysical and spiritual experiences he had had in Auwahi. "That stuff dishonors truth. But when something is real, it's real." He smiled, then admitted in a mock-confessional tone that he was in fact a tree hugger. "Have you ever actually tried it?" he asked. "You can blindfold me and I can always tell a live tree from a dead one just by hugging it—I can feel the life force. I've got some pretty hardened people to do it, and now they too have literally become tree huggers themselves."

Of course, not everyone gets Art. One veteran member of the

Hawaiian conservation community put it this way: "I'm just not into all that huggy-poetry stuff. I'm not interested in getting in a circle and holding hands and shaking leaves at each other—I can't do that kind of stuff without feeling silly. Everything Art does has to be intensely spiritual, but he's so sincere, and so knowledgeable, that of course you have to respect what he's doing."

Nevertheless, most people seem to love him to death and are charmed by his knowledge and charisma. One local woman who has known Art forever said simply, "Art impassions people. Sure, he is passionate about the environment and Hawai'i's native species. But he also cares deeply about the local people, the Hawaiian culture, and especially the Hawaiian kids. Not very many people who do what he does care about those things too."

Another volunteer summed up her experiences of being at Auwahi with Art as "magical and unbelievably beautiful." She continued, "He often begins by just asking us to listen to the silence and smell the mountain air. Sometimes we'll do chants to reintroduce the people and plants, and everyone will get chicken skin. Each time he'll tell us a different story, but it's always about Auwahi, so we try to envision it as it was, and how we got to where we are today, and what we're going to do today to change things, and everybody just gets totally inspired."

Commenting on this, Art told me,

> Sure, all that stuff is really important to them, but they also have a huge effect on me. To have forty people come out, most of whom I've never even seen before, and be so enthralled with what I have to say and what we're doing up there—it's an amazing thing for a scientist to experience. They come because they really want to help the forest; they want to do something real. In many cases they have highly tuned bullshit detectors—they look me in the eyes, weighing it all. They'll start out pretty cynical, wondering how long they have to work and when they get to leave. You can't convert them all, but by the end of the day, for most of them, their cynicism is broken, and they end up sounding like kids in their optimism. They'll say things like, "Hey, we can't leave yet, we haven't finished weeding this section," and "If

we could do this every day, we could do hundreds of acres!" and "Now we don't have to go to church tomorrow."

Art downshifted and slowed the truck to a crawl to navigate the even rougher, deeply rutted last two miles of the road to the ten-acre Auwahi exclosure. On my previous trip in 1999, Art explained that he had chosen to start working within this particular area because of its high number of endangered species, high density and diversity of native tree species, and its relatively high abundance of remnant native shrubby and herbaceous species. He had become convinced that Auwahi's dry forests had once possessed a thick understory that provided a critically important microenvironment for the germination and establishment of the larger native flora.

Constructing a durable fence in this remote, steep, and rocky terrain that would reliably exclude the region's cattle and feral pigs had been immensely challenging. To accomplish this project, in the summer of 1996 Art had assembled a broad coalition of volunteers that included about fifty local hunters and community members to tackle the first fence-building tasks, such as designing the exclosure, acquiring posts and other materials, hauling supplies and equipment to the site by hand and helicopter, and digging and cementing in the fence poles. The actual fence-building steps—rock drilling to pound in the fence pins, clearing the brush away from the fence lines, and stretching and attaching the hog wire—were finally completed in the spring of 1997 with the assistance of professional work crews from Haleakalā National Park, East Maui Watershed, Maui Land and Pineapple, and the Biological Resources Division of the US Geological Survey. To the best of Art's considerable knowledge, their fence protected a tract of Auwahi dry forest from the ravages of ungulate browsing and trampling for the first time in hundreds of years.

Art's next step was to go after the thick, smothering layer of kikuyu grass that dominated the understory within the fence. After several failed attempts due to gusty winds, a highly skilled helicopter pilot finally managed to fly in and maneuver a ball spray rig in and around the native trees and shrubs and spray about 75 percent of the grass within the exclosure.

Just as we had experienced on the Big Island after controlling the fountain grass within our restoration projects, after they killed the kikuyu grass this exclosure was quickly invaded by a suite of formerly suppressed alien weeds that included a dense stand of the same species of milkweed (*Asclepias physocarpa* or balloon weed) we had struggled to contain. Like us, Art and his volunteers used a combination of chemical and manual control techniques to combat these new weeds while they ramped up their native seed collection, greenhouse propagation, and outplanting operations.

On my previous visit, Art had pointed out several instances of "natural" native seedling recruitment within the Auwahi exclosure, and he explained that these were probably the first such examples for some of these species on East Maui in the last fifty to one hundred years. He was especially excited about discovering a single *hala pepe* (*Pleomele auwahiensis*) seedling growing within the partial shade and accumulating leaf litter of its presumed parents. This culturally important member of the agave family is found only on Moloka'i and Maui and is one of the most common native tree species of Auwahi and the other remnant dry forests on the island. Yet after we had all knelt down and examined it, Art said that it was the first wild *hala pepe* seedling he had ever seen in his nearly two decades of intensive botanical work on leeward Haleakalā.

Despite these kinds of exciting and inspiring phenomena, however, the future ecological trajectory of the exclosure at that time was anything but clear. Would they be forced to perpetually "garden" that plot forever? Would the weeds take over the minute they stopped killing them? Was the vision of a "self-sustaining" native dry forest within that Auwahi exclosure, let alone the larger surrounding area, a noble and attainable goal or a grand delusion?

Just as I had felt in many other restoration projects, my intellectual point of view on these kinds of questions at Auwahi had depended on my physical point of view: when I was looking at an impressive patch of native plant regeneration, anything seemed possible. But when I was looking at the remaining swaths of kikuyu grass in the gullies, the suites of other weeds racing into the recently cleared areas, and the vast degraded and tattered landscape surrounding the exclosure, restoring this landscape had seemed almost laughable.

Art finally pulled off the road and parked the truck above the

Volunteers weeding in East Maui, 2002. *Forest and Kim Starr*

Volunteers preparing Auwahi One for restoration, 2002. *Erica von Allmen*

ten-acre exclosure he calls "Auwahi One." I got out and drank in the cool, misty air. While it was difficult, time consuming, and expensive to transport people and supplies to this site, Auwahi clearly had its advantages. Despite its tragic ecological history, the breathtaking landscape, panoramic views, and deep, ethereal stillness created the ironic illusion of being in the middle of a vast, untrammeled wilderness. At an elevation of four thousand feet, its climate was cool, comfortable, and often quite close to the "Paradise of the Pacific" fantasy that most tourists mistakenly think is true of the Hawaiian Islands as a whole. The combination of this benign climate, deep pockets of rich soil, and location above the reach of some of the archipelagoes' worst alien pests also made Auwahi an excellent place to grow native plants.

I looked over in the direction of the exclosure and was stunned. Even though I couldn't make out the actual fence lines from that distance, I couldn't miss the large patch of green leaping out of the surrounding landscape like a picture in a child's pop-up book. Art looked at me and beamed. "I deliberately made the exclosure square so that people wouldn't mistake it for some kind of natural feature. You can

Auwahi One fence line, 2007. *Erica von Allmen*

actually spot it now when you're flying over this part of the island. That's one measure of success I thought I'd never live to see" (see Plate 11).

As we walked down to the exclosure, Art explained that throughout his many years of observations, Auwahi's native trees tended to be pale and ragged looking. Yet after they built the fence and removed the kikuyu grass, the trees became greener and more vibrant than he had ever seen before. Since the trees outside the exclosure still looked the same, he had concluded that this change was due to the absence of the grass and the ungulates.

When we entered the exclosure, I immediately became disoriented. Whole sections that were still smothered by kikuyu and other weeds on my last visit were now dominated by natives. In fact, an impressive abundance and diversity of regenerating Hawaiian vines, shrubs, and trees seemed to be everywhere. "Look!" Art said, pointing to a pile of leaves that had accumulated at the base of a large native canopy tree: "It's a real forest."

He explained that shortly after my last visit, they had hand sown over a million 'a'ali'i (*Dodonaea viscosa,* a relatively common and

Auwahi One, 2012. *Erica von Allmen*

hardy indigenous shrub in the soapberry family) seeds to "occupy the beachhead" created by the removal of the kikuyu, help control the newly emerging stands of weeds, and create favorable microhabitats for the reestablishment and spread of more native plants and arthropods. Over the years they had also outplanted thousands of individual seedlings and saplings of rare and common native plants. As we walked around, in some places there was such a thick mass of vibrant young native plants that it was difficult for even Art to tell which had been seeded, which were transplanted, and which had come up on their own. When I pointed to a clump of mature, fruiting, flawless *'a'ali'i* shrubs and inquired about their origins, Art just shrugged. "I don't know," he said. "Many of our babies have been having their own babies for quite some time now."

When I told Art about our struggles on the Big Island to figure out what the right restoration mix of native species should be, he just threw up his hands and shrugged:

> It's unknowable. But the truth is, I don't even care anymore about what the "correct" proportion of each different species might be—I'm fighting a war! The purpose of my dry forest work here is to reestablish regimes of competition between native species; I'm willing to trust the balance that emerges from that process. Even if I wanted to, which I don't, there's no way I could micromanage and control that balance anyway. You know, species like koa and *'ōhi'a* used to dominate these islands and function as each other's major competitors. But now, most of the time, we're lucky if there's even one *'ōhia* and one koa remaining in the same pasture, and you can almost hear them calling out to each other: "Hey *'ōhi'a lehua,* brah, you OK out there?" "I'm hanging on, koa, I'm hanging on."

Art explained that part of his vision was to one day be able to responsibly extract various materials from the forests of Auwahi, make things out of them, and maybe even sell some of these products to help support the restoration program:

Sure, I want to have preserves where trees rot and nothing is taken. But I'd also like to have other places where we use the forest. You know, the ocean was the Hawaiians' refrigerator, but these dry forests were their toolboxes and medicine chests. Take 'a'ali'i, for example. They used its beautiful strong wood for house construction, weapons, and farming and fishing tools. They used its leaves for medicines. And they made leis and a bright red dye out of its fruit capsules. In fact, of the fifty native tree species remaining out here in Auwahi, we know that the Hawaiians used about 40 percent of them for medicine, more than a quarter for making tools, canoes, and houses, and a handful for miscellaneous things ranging from food to firewood to fireworks. Only nine of the trees have no recorded uses, but because the wood of most of those species appears to be of good quality, I bet they probably used them for things that have long since been forgotten.

He grew increasingly animated as we wound our way down deeper into the exclosure. Walking through the "forest" with him was like walking with an expert curator through a natural history museum in which all the specimens had just broken out of their display cases and come to life. "Look at this vignette!" he'd say, pointing to something he found especially interesting.

"Check out the color of this po'olā's [*Claoxylon sandwicense*, a member of the spurge family] pollen!" he cried, thrusting it under my nose. "It's yellow now, but when it dries, it turns into an unbelievably beautiful blue.

"That's the remains of a long dead *Partulina* snail," he said, pointing to a half-inch long bleached shell lying in the leaf litter beneath a large native tree. "Their presence here indicates that at one time this must have been a much wetter and more densely forested site than it is today.

"Listen to those 'amakihi! I've been hearing a lot more of them since we started the restoration—they're even nesting in here now.

"Take a whiff of these," he said, handing me a bunch of colorful flowers with a strong but pleasant citrus scent.

And that sweet, honeylike odor that's been following us around? That's the maile [*Alyxia oliviformis*, a unique vine in the dogbane family]. Probably no one has experienced this combination of colors and smells for hundreds of years, or maybe ever, given all of the changes that have occurred since these plants and people coexisted. That's another goal of mine—to restore the colors and perfumes that have been lost from this landscape.

Look at that amazing diversity of lichens growing on those trees! The lichenologists tell me that the Auwahi flora is one of the richest, maybe *the* richest, in the world. Even though they've barely explored this place, they've already collected over two hundred lichen species in sixty to seventy genera; nowhere else is even close to that in the Hawaiian Islands, and those numbers compare favorably to the world's other known lichen hotspots. You know all those new species of spiders we've been finding in here? In turns out that many of them are cryptically colored and apparently live in the lichen.

Art told me that he also remembered the faces of all the people who came out for the workdays, what each and every gully looked like before and after their work, and even what they planted. He pointed to a clump of *hala pepe* that were all about ten feet tall and laughed. "One day a bunch of us ate lunch there, and then afterwards we decided to immortalize that spot by each planting a *hala pepe* tree. They were just seedlings then, but now look at them—every single one of them has thrived! I've been out here at night and seen native moths pollinating some of the *hala pepe*s within this exclosure."

A few minutes later, Art put his hand around the trunk of a vigorous, twenty-foot-tall tree and gave it an affectionate little shake. "Recognize this guy?" he asked. Although its compound, leathery leaves looked vaguely familiar, I had no clue what it was. "This is one of those *māhoe* you and Lyman brought over from the Big Island way back when." I looked up at the tree again, dumbfounded.

A few minutes later, it was Art's turn to get chicken skin. "Holy shit!" he said, pointing to a clump of fruits on a sprawling, viney-looking shrub. "This is the first time in fifteen years that anyone has ever seen *Melicope adscendens* in fruit. Look at those four beautiful separate carpels." He held up his hand for a high-five. "Holy shit!!!"

I was familiar with this genus in the rue family because it is among the largest of all the flowering plant genera in Hawai'i. All forty-seven of the species listed in the *Manual of the Flowering Plants of Hawai'i* are unique to the islands, many of them are federally endangered, and two are known only from old herbarium collections and thus may already be extinct. (The authors of this manual actually placed these species in the genus *Pelea* because the submersion of this taxon into *Melicope* was still in progress at the time of the *Manual*'s publication.)

Once he finally calmed down, Art explained that this *Melicope* species was first discovered at Auwahi in 1919 by a botanist from the Bishop Museum in Honolulu. But it was not seen again until Art and another colleague rediscovered a few individuals in western Auwahi in 1981. Over the course of their subsequent surveys, they found about thirty-five more individuals, but over time the vigor of those plants steadily deteriorated, several of them died, and those that managed to hang on apparently no longer had enough energy to produce fruits and seeds.

"This exclosure is so pollinator friendly, these fruits could easily have been produced by outcrosses with some of the other *Melicopes* in here, although of course they could have been self-pollinated as well," Art concluded as he continued to stare incredulously at the fruits. "I'll definitely be taking the volunteers here tomorrow—I love to show them these little victories because there have been far too many sad tours of Auwahi. I think at some point people got too comfortable with all that sadness, and they eventually just gave up."

We finally reached the bottom and set off through the thick, unmanaged kikuyu toward the second Auwahi exclosure. After a while, Art turned to me and said, "Notice how quiet we've become? The same thing always happens with the volunteers out here. The whole time we're working in the exclosure, people are talking and laughing and full of energy and hope and good spirits. Then when we come out here, into these abused lands, things suddenly get real quiet, and everyone seems to be lost in their own private thoughts."

When we reached the entrance to "Auwahi Two," Art tenderly kicked the fence.

Did I tell you the story behind the creation of this baby? One night one of the 'Ulupalakua Ranch guys won big in a poker game. He called me up and said, "Art, get your ass out here and flag out the dozer line for the second exclosure!" I ended up marking out this twenty acres right here. If I had had to pay for it, that dozer would have cost me $9,000 a day. But you know what? Some of these dozer drivers have become my heroes. They tell me, "We're always considered the bad guys, but we want to help too." I've worked with them to put in some extremely difficult and dangerous conservation access roads and exclosures in other areas of leeward Haleakalā in which everyone said it couldn't be done. The courage, skill, and tenderness of these guys was just mind-blowing—those days contained some of the most profound and spiritual moments of my life. And a lot of them are really interested in the plants, too—they want to learn their names, they want to hear the stories, they want them to survive. I used to think the sound of conservation was the "dink-dink-dink" of our shovels glancing off the lava, but now I think it's also the growl of the dozer engines and the "ding-ding-ding" of their backup warnings.

We entered the second exclosure and surveyed the progress of the restoration work they had only recently begun at this site. There was still a lot of kikuyu and other weeds around and only a few small, scattered patches of native plants that were just beginning to establish. Nevertheless, we saw an encouraging number of native seedlings popping up on their own in the shade of the exclosure's remnant trees and shrubs.

"This time we are spraying the grass from the ground only," Art said, "and we are doing it slowly and methodically so we don't create any large openings for the weeds before we are ready to come in and fill the gaps with our seeds and plants. See that?" he asked, pointing to a clump of milkweeds spreading across a large section of dead and decaying grass. "That's too much, too fast!" We walked over and pulled some weeds that were emerging in their recently dug outplanting holes. "It's hard to hold the volunteers back—sometimes they get so excited that they just keep going without thinking about the bigger picture and the coordination of the whole operation."

We came to another *Melicope, M. knudsenii*, that I knew from my previous visit was the very last individual in the wild and one of the reasons why Art had chosen this area for his second exclosure. "That's where the cattle and pigs used to rub up against it," he said, pointing to an old scar running across its trunk. He was silent for a while as he touched the tree and looked up into its branches.

I never thought I'd live to see that beautiful folded leaf again. When I first started doing botanical surveys up here, there was something like thirty of these still left, but with each successive visit there'd be fewer, until there was just five, then four, three, two, and then just this one. The knowledge weighed heavily on me—I'd wake up at two in the morning, wondering what I should do. Finally I decided to start telling other people about it and share the responsibility.

Then this one started looking bad. I saw some Argentine ants on the stem, which is always bad news. But since we fenced this area and sprayed out the grasses, it's been doing much, much better. And the *Melicope knusendii* we've outplanted in here are doing great too.

Although they had never been able to get any viable seeds off that last remaining wild specimen, they were able to germinate seeds from an old cultivated tree that had been dug up from Auwahi and transplanted as a seedling in the 1920s. They eventually transplanted about ten of the plants produced from those seeds near the original mother tree; today all of these are healthy, and the largest is about six feet tall.

We walked over to another section of the exclosure in which an army of knee-high 'a'ali'i shrubs was dominating the understory. "Remember this," Art said, spreading his hands wide, "because you won't recognize it next time you're here. We've basically weed proofed the first exclosure now, and we'll do the same thing down here. Only this

Auwahi Two before fencing, 2003.
Forest and Kim Starr

time, by applying what we learned in Auwahi One, I'm confident we can do it for about one-fifth the cost, and about five times faster, than we did up there. I know what to do now, and how to do it. I've got a great group of dedicated volunteers. And being able to finally hire Erica von Allmen . . ." He smiled, searching for words. "It's critically important to be surrounded by people on the ground who really care, and within the sea of all the great and caring people I've worked with, she's simply the best. Her heart is huge, and humble."

Having barely survived the fast, furious, and far-ranging experience of being with him in the field before, I had sought out Erica for some basic background information prior to my Auwahi tour with Art. But when I asked her for a slow, comprehensive overview of their scientific research and ecological restoration program, she just laughed:

THE PŪ'OLĒ'OLĒ BLOWS

Auwahi Two, 2012. *Fernando Juan*

We used to be a lot more structured in the beginning—we had our little trials and more formal experiments—sun versus shade, seedlings inoculated with mycorrhizae versus no inoculation—but as we learned more, it eventually just became largely a matter of getting as many plants in the ground as possible. Once they are in the ground, we don't even bother to water, tag, or track them anymore unless we're working with something that's really rare. I'm sure it'd be valuable and fun to do more monitoring and experimentation, but we're always just too busy, and all that other stuff takes so much time and effort and money.

I told Erica about my experiences attempting to do restoration by committee on the Big Island and asked her whether they ever had to work their way through the inevitable conflicts that emerge when people have different visions of what to do and how to do it. "Well, Art is his own entity," she said, laughing again. "We don't have any formal management plans or visioning documents for Auwahi. He constantly assesses the situation up there and then makes his decisions accordingly. He knows what he is doing, and everyone respects that." Erica continued,

> In the beginning, maybe one or two people would show up for our volunteer days. We'd be like, "Alright, somebody came!" But now it's a zoo—we're shuffling vehicles and plants and people and equipment, struggling to contain the chaos. But once we get up there, we really just turn everyone loose and let them plant. You stand next to some of the trees now and see all these little rocks around their bases, and then you realize the trees have pushed those rocks out from the little circle we made around their trunk when we planted them.

When I asked Erica if there were any concrete lessons she had learned through her work in Auwahi, she had to stop and think about it for a few minutes. "Well," she said finally,

> I do see different levels of complexity that I didn't see before. For example, are we imposing an artificial selection regime by always selecting the plants from the fastest germinating seeds in the greenhouse? What's happening to the genetic diversity of our rare species and our dry forest flora as a whole? Will some of the species that appear to be doing so well now eventually not make it due to the rats, or the loss of their original pollinators, or something else that we haven't thought of or can't foresee? Because there are so many interacting factors that are virtually impossible to track and tease apart and comprehend, I guess I've just learned to accept the limitations of our knowledge and believe that in the end, the mountain knows best. Of course, because our time and resources are always stretched to the breaking point, we don't have the time to focus on those kinds of questions anyway.

But I feel blessed to be a part of this project. Every now and then, I think about going back to college and studying ecological restoration more formally, but then I realize that no school could possibly be as good as working with Art in Auwahi.

(Erica did eventually go back to school on the mainland, but after earning a masters degree in biology in 2012, she moved back to Maui and resumed working with Art at Auwahi.)

Art and I walked back up to the truck, retrieved our lunches, and sat down on a flat rock with an expansive view of the exclosures and the slate-blue ocean below. He stomped the boulder in mock anger and explained that there really shouldn't be any bare rocks up here because everything should be covered over in vegetation.

Eventually, our conversation turned toward the broader implications and significance of Auwahi and the ecological future of Maui and the Hawaiian Islands as a whole. He recalled how as a local boy growing up on Oʻahu, he often went into the forest to get away from people. At first he was only interested in the species' Hawaiian names and their cultural stories, but when he started to learn about the troubles of the native flora and fauna, he became motivated to learn their scientific names and deal with all the people and institutions involved with Hawaiian conservation.

We talked about how sad it was to watch the ongoing deterioration of so much of Hawaiʻi's native biodiversity and the uphill struggle of getting the public and the politicians to understand the consequences of this tragedy and care enough to do something about it.

"All that is sad," Art concluded, "but it doesn't discourage me. What does discourage me is the way the people who are supposed to be doing something about it clash with each other and that so many of them have given up hope. That loss of hope is really our biggest enemy. When we say it's not possible—when it becomes just a paycheck—when we believe it's not worth it for my generation or future generations to even explore what is possible . . ." His voice trailed off. "I have had the greatest of *kumu*. They did critically important and heroic things during their time, but now it is up to me, to us, to our generation, to carry their work forward."

When I asked Art what lessons he had learned at Auwahi and elsewhere over his many years in the trenches, he immediately said that solving the ecological problems is actually less difficult than solving the human problem:

> Successful conservation requires more than good science and hard work. It requires a strong human presence. Some might naively think that after the scientists reveal their eloquent work, the public will say, "Oh, thank you, thank you—now we're ready to implement your ideas!" You and I know it doesn't work that way. I've become increasingly interested in developing and sharing knowledge with an ever-broader community of people. When we reach the tipping point of enough people with firsthand knowledge getting involved, people who have seen with their own eyes and done with their own hands, really good things can happen.

When I asked him how he decides what to do and how to do it, he said that he mostly got his ideas directly from ecological observations. He explained that this was much easier back in the 1980s when he just did biological surveys and had more time to look closely at what was happening in the field. Referring to Maui's "Grand Master of Native Plants," Art went on:

> And then of course, there's Rene Silva's saying, which has become a foundation for an awful lot of what I do: "Mo' betta we try something, 'cause we already know what happens when we try nothing." The first time I heard that I thought, "Wow!" I've been taught the scientific method, which is accurate but conservative and doesn't encourage one to go out on a limb. Now I try to follow Rene's method, and I try to have courage, but big ears, too. But you know, when I first started talking about some of my really big ideas for how to conserve and restore Maui on a landscape scale, I used to be laughed out of the room. But now the landowners, the people in the governmental organizations, the funding agencies, the business community—when I talk about that stuff now,

they all have this serious look on their face, start nodding, and want to know more.

On Maui and across the other Hawaiian Islands, perhaps the most important entities for articulating and implementing the kinds of big ideas Art and other conservationists have are the so-called watershed partnerships. For example, the East Maui Watershed Partnership was formally established in 1991 through a nonbinding agreement signed by the County of Maui, the State of Hawai'i, the Nature Conservancy, Haleakalā National Park, East Maui Irrigation Company, and two major ranches. As Alex Michailidis, East Maui's Watershed coordinator, told me, "They all signed the agreement to protect this region's native forests because they all realized that that was where their water comes from: no forest, no rain, no water!"

Alex explained that while the separate organizations within this partnership did some good things on their own, they eventually realized that to make real progress they needed someone to coordinate the entire operation, get funding, and hire the staff necessary to get the job done. All of the people involved with conservation on Maui I spoke with had nothing but praise for this partnership. As one senior member put it, "This watershed partnership is the best thing that ever happened to conservation on Maui—99 percent of the work that's getting done here is through this partnership. We come together, get out the maps, and say, 'Hey, wouldn't it be nice to put up a big fence all across there?' And then the partnership finds the money and does the work, and it actually gets done."

Alex told me that most of what his crews do is build fences, remove ungulates, and try to control the major weeds, and he said that most of their money comes from grants, Maui County, the State of Hawai'i, and various private foundations. When I mentioned the perpetual tensions on the Big Island between the conservation community and the hunters and landowners, he said that things were much better on Maui:

> Here, the private landowners are mostly onboard because they generally don't have the money to manage their own watersheds anyway, and they realize they can get a lot more

done by working together and leveraging the staff and money we can get through the partnership. They also are beginning to understand that they need to take care of their upland forests to ensure that they will have enough water for their own needs in the future. However, while they get these kinds of selfish things out of it, plus some very valuable good PR, at least some of them are in this because they are also good guys and they know it's the right thing to do.

Alex explained that while there were a small number of very loud hunters that opposed what they were doing, it was nothing like the Big Island. This difference may be at least partially due to the fact that unlike the Big Island, the only way to access most of Maui's high country is by helicopter, and thus very few people ever hunt up there anyway. Nevertheless, to appease the hunting community, they sometimes flew hunters up after their fences were complete and let them take the first shots at the pigs.

They love it—they take movies up there and come back with some great stories. Of course, at $700 per hour of helicopter time, it's an expensive way to get your meat—the price tag can easily come out to about $5,000 for fifty pounds of pork! But it's great PR for us, and hopefully it generates some goodwill. You know, both the hunters and the conservationists share a passion for Hawai'i's environment—both sides appreciate its beauty and have respect for it. I don't think we're ever going to convince some of them that the pigs are bad, but when you bring them to those relatively pristine forests without many weeds, they feel that power deep down inside too, and then at least some of them say, "Wow, this is what a native forest looks like. We really should help protect these areas!"

Alex and many of the other members of Maui's conservation community I spoke with also stressed the importance of hiring local guys to work on the watershed partnership's crews. Over the course of their work, many of these guys have become strong believers in the overall cause of conservation and have gone on to inform and "convert" their peer groups far more effectively than the haole conservation community ever could. In fact, several people who I consider to be extremely dedicated and skilled workers themselves told me that Alex had created the most hardworking and efficient crews they have ever seen.

Art has also become increasingly involved in another similar group called the Leeward Haleakalā Watershed Restoration Partnership. Through this program, he hopes to realize some of his many other dreams for this part of the island, including extensive multimillion-dollar fencing projects, rare bird reintroduction programs, and massive koa reforestation trials. Pointing down toward the exclosures, he said,

> Look, we've made a great start at Auwahi. We're getting closer to the day when we can essentially walk away from these "restored" forests. I see these not as gardening projects but rather as developing technology to save ecosystems. We learn how to repair the ecological fabric, then progressively start isolating the factors to figure out what's limiting when things fail: Is it rats? Pollinators? Seed dispersers? Genetics? Then we start ramping up what we learn in places like this to a landscape scale.

"Ramping up" is exactly what Art and his colleagues have done at Auwahi since we had that conversation in 2005. "We've also dramatically refined our restoration techniques," Erica told me in 2012:

> For example, we now spray the kikuyu grass in the area we're going to plant about a month ahead of time. This allows us to use the thick layer of dead and dying grass as a weed barrier and mulch. In the year or so it takes for all that kikuyu to decompose, the fast-growing natives such as ʻaʻaliʻi that we outplanted in there have established themselves and begun to drop their own leaf litter. Consequently, our overall weed problems have declined dramatically, and we are seeing incredible levels of native plant reproduction, including some

endangered species that had not regenerated on their own in who knows how long.

Using these and other new and improved methodologies, Art's crew and volunteers are now able to routinely outplant fifteen hundred to two thousand native seedlings in a single day (in the early 2000s, they considered transplanting a few hundred plants to be a big day). As of January 2012, their volunteers had logged a grand total of 27,065 hours and outplanted 88,697 native plants. In the last few years, they've also moved beyond Auwahi Two and planted twelve acres within Auwahi Three, and they have enclosed all three of these restoration projects within an overarching 151-acre exclosure that protects most of eastern Auwahi.

The last time I saw Art, he was even more excited about the future ecological possibilities within and beyond Auwahi than he had been

Recently fenced section of eastern Auwahi awaiting future restoration. *Fernando Juan*

back in 2005. He rattled off numerous examples of the kinds of ever more ambitious restoration projects he argued could and should be done on leeward Maui and across the Hawaiian Islands. But then, after listening to himself go like a runaway train, he suddenly stopped, took a deep breath, and smiled. "You know," he finally said, "W. S. Merwin [the famous Maui poet and recent poet laureate of the United States] told me, 'Art, you have one job for the rest of your life: you have to tell the story, make the story better, make yourself cry at the right time, make them cry, and share the knowledge.'"

8 TURNING HANDS
Limahuli Botanical Garden, Kaua'i

Although all my filthy clothes were still spinning in the washing machine, I couldn't stay inside any longer. The night was too inviting; the gurgling stream beneath my window too tantalizing; the moonlight filtering down through the branches of the swaying breadfruit trees too hypnotic. "What the heck," I finally told myself as I tentatively stepped outside stark naked, "I'll never get a chance to do this again."

Outside, the heavy sea air was warm, moist, and salty. I stood on the steps, listening to the frogs croak and making sure the coast was clear. Satisfied, I stepped off the lanai and walked up the path past the National Tropical Botanical Garden's Limahuli Visitor's Center, then stopped beside a trickling canal that wound its way through a series of restored seven-hundred-year-old stone walls that terraced this sloping landscape into a series of irrigated taro patches, or *lo'i*. The underground, potato-like stems of this plant were a staple food of the ancient Hawaiians, who eventually named and cultivated more than three hundred different varieties. Even today, just up the road from the garden, much of the broad wet valley opposite Kaua'i's Hanalei Bay is devoted to growing large quantities of taro to supply the islands with fresh poi I couldn't make out the tops of the knife-edge ridges that rise some two thousand feet above both sides of the valley, but I knew that the precariously perched Pōhaku-o-Kāne (stone of Kāne) was sitting somewhere up there off to the east. According to Hawaiian legend, this rock, which rolled across the ocean floor from Tahiti to get here, was honoring his promise to stay awake and watch and remember all that went on below until the great god Kāne returned.

I looked up to my right and thought about the history of the spectacular Makana Mountain that towers majestically over Limahuli Valley's west side. This site, popularized as "Bali Hai" in the movie *South Pacific,* is one of only two known locations in the islands where the famous fire-throwing ceremonies were held. On very special occasions, a group of highly trained fire throwers would load up with bundles of dried *pāpala* wood (*Charpentiera elliptica,* a small tree in the amaranth family that is found only on Kaua'i) and scale the cliffs until they reached the mountain's pinnacle. When it was dark, they set the logs on fire and hurled them out toward the ocean. The updrafts created by the trade winds slamming against the cliffs of Makana kept the light, hollowed-out *pāpala* aloft and sent the flaming sticks as far as a mile out to sea. The Hawaiians watched these fireworks from the land and in their canoes at sea, and some tried to catch the firebrands to demonstrate their affection for a special loved one.

The last time this ceremony was attempted, early in the twentieth century, the fire throwers apparently couldn't find enough *pāpala* and decided to use *hau* (*Hibiscus tilliaceus,* a small tree in the mallow family that also produces very light wood; this species is widely distributed in coastal and riparian areas throughout the tropics and subtropics; whether it reached these islands on its own or was brought over by the Polynesians remains unresolved). Unfortunately, however, several unplanned terrestrial fires erupted after the wind blew the burning *hau* back onto the land.

Given Limahuli Valley's sheltering mountains, perennially flowing stream, frequent rains, fertile soils, and accessible, productive marine ecosystem, no one was surprised when recent archaeological evidence corroborated the local belief that this area was home to one of the earliest Hawaiian settlements. As I admired the restored *loʻi,* I tried to imagine this landscape in its primeval condition, before the arrival of the weeds and ungulates and rodents and mosquitoes. Then I tried to picture the prehistoric days, when this valley was thickly settled with Hawaiians subsisting on a rich diet of seafood, taro, breadfruit, banana, and dozens of other nutritious plants and animals. It was hard not to romanticize that world—a tightly knit clan of wise, peaceful people living out their simple, bucolic lives in a tropical paradise hidden from the rest of the troubled planet. Even though I knew the truth was far more complicated, at that moment it seemed impossible to believe that such things as shark-toothed clubs, highly stratified societies, and strict *kapu*s were a very real part of that world as well.

As I tried to visualize what this valley might look like one hundred years from now, I thought about a mind-bending passage I had read earlier that night in a book of local oral history interviews by Kepā and Onaona Maly (*"Hana ka lima, ʻai ka waha"*). In a section titled "The Not So Distant Past," the researchers who had conducted those interviews cited the following quote from another book about Hawaiian history:

> It is interesting to note that in Hawaiian, the past is referred to as *Ka wa ma ua,* "the time in the front or before." Whereas, the future, when thought of at all, is *Ka wa ma hope,* or "the time which comes after or behind." It is as if the Hawaiian stands firmly in the present with his back to the future, and his eyes fixed upon the past, seeking historical answers for present-day dilemmas. Such an orientation is to the Hawaiian an eminently practical one, for the future is always unknown, whereas the past is rich in glory and knowledge. It also bestows on us a natural propensity for the study of history.

Was the past the key to the future of the Hawaiian Islands? Could it serve as the glue that somehow might hold all of this garden's disparate present operations together? I looked up in the direction of Pōhaku-o-Kāne and thought about all the things he must have seen since Kāne first lifted him up and placed him on that mountain top, long before the first people ever discovered these islands. If, like the ancient Hawaiian kahuna, I could understand his language, what would he have to say about the strange activities that were presently unfolding beneath his watchful gaze?

I woke up early the next morning, ate a quick breakfast, slathered myself in sunscreen and mosquito repellent, and walked (fully clothed) up into the garden to meet Matthew Notch, Limahuli's restoration project manager and volunteer coordinator. Matt was leading a Sierra Club service project/ecotourism hike that morning and had kindly invited me to tag along. I was a few minutes early, so I ducked into the garden's Visitor Center. Unlike most of Hawai'i's other tourist attractions, this modest building (actually a ten-by-thirty-two-foot modular office trailer that they had skillfully transformed to resemble a historic, plantation-style home) offered its visitors insightful and substantive information and tasteful, reasonably priced, often locally produced items for sale. I was also pleased to see that Phyllis Somers and Nancy Merrill were working behind the counter. Both of these women had been volunteering and working at Limahuli for many years, believed deeply in the garden's mission, and were exceptionally good at interacting with and engaging their many visitors.

"Aloha, Bob!" they both said at once, giving me a hug. "You won't believe how things have changed up in the valley since you were here last," Nancy said proudly. "I tell everyone I know who thinks that restoration doesn't work to go up there and take a look. It's just impossible to come out of there feeling anything but optimistic."

The three of us chatted until I saw a small crowd assembling around Matt in the parking lot. When I joined them, he was informing his dozen volunteers that the Limahuli Garden and Preserve consisted of approximately one thousand acres, with three distinct subdivisions: the lowland montane rain forest of the Upper Preserve; the mesophytic and lowland rain forests of the Lower Preserve; and the more formal, seventeen-acre, publicly accessible Botanical Garden at the mouth of the Lower Preserve in which we were standing.

He then pointed to the sculpted peaks of the Upper Preserve off in the distance and explained that that relatively intact, four-hundred-acre hanging valley goes from about 1,600 feet to 3,330 feet in elevation and can be accessed only by helicopter (see Plate 12). The Limahuli Stream, one of the most pristine in the entire state, plunges nearly eight hundred feet in a spectacular waterfall from the Upper Preserve down to the back of the Lower Preserve, then winds its way through this six-hundred-acre, bowl-shaped lower valley to the sea.

Matt then highlighted some of the numerous activities currently underway at Limahuli: the efforts to preserve and restore the upper valley's rain forest that had been devastated by Hurricane Iniki in 1992; the research and outplanting programs within the degraded mesic forest of the Lower Preserve; the long-term Limahuli Stream research and monitoring project; the archaeological investigations

Portion of Limahuli's restored *lo'i*, or taro patch.
National Tropical Botanical Garden

into the settlement and use of the Limahuli Complex; and the various outreach efforts of their Visitor Program. He told us that after we finished our work within one of the outplanting sites in the mesic forest, we would hike to the back of the Lower Preserve to see the waterfall and the recently completed *hau* removal project.

Matt concluded his introductory spiel by explaining that the entire Limahuli Valley was located within the Limahuli Valley Special Subzone of the State of Hawai'i's Conservation District, which was governed by the Department of Land and Natural Resources. Mrs. Juliet Rice Wichman donated the first thirteen acres of Limahuli to the National Tropical Botanical Garden in 1976, and the rest of the thousand acres had been gifted by her grandson, Charles "Chipper" Wichman in 1994.

A visitor on Limahuli's self-guided tour. *Robert Cabin*

A few nights earlier, I had heard more of the story behind the creation of Limahuli Garden over the course of a long evening of doubles ping-pong on Kaua'i's south shore with Chipper, his wife Haleakahauoli, and Richard Hannah, the garden's librarian. Chipper began:

> Even though I grew up here, like almost everyone else, I really wasn't aware of what was going on out there in the field ecologically. I just knew it was all green and beautiful, and the stream was cold and clear and great for swimming. I assumed that the stream had always been choked by the *hau,* and the bamboo forest back in the Lower Preserve had always been there. Later, when I was searching for some direction to my life as a teenager, I spent half a year working at Limahuli and learned about my grandmother's vision to turn it into a botanical garden. Although she wasn't especially interested in native plants, she was into Hawaiian concepts of land preservation and stewardship, and thus she wanted to turn the valley into a place where people could learn more about Hawai'i and why this place is so special

Following his grandmother's suggestion, Chipper worked as an intern in the National Tropical Botanical Garden's Allerton Garden on the south shore, then started doing exploratory botanical surveys of Limahuli in 1976. He recalled,

> We would send our specimens off to Dr. Harold St. John at the Bishop Museum in Honolulu, and as you know, St. John is the world's greatest splitter—according to him, we discovered seventeen new species! I know most botanists think he's crazy, but he told me later that one reason he went overboard on the splitting was that he believed that giving the plants unique names helped protect them—if you lump them all together, people might think, "So what if the Limahuli population goes extinct, we've got others." Of course, today we consider that kind of fine-scale variation to be the beginning of divergent evolution, and consequently most of our "species" have been sunk into synonymy. But I still

have the letters St. John sent us in which he gushed over our "fabulous finds" that got me all fired up. It was great for my grandmother too, because she always believed that Limahuli's plants and culture were special. Who knows, maybe St. John was really onto something and was just ahead of his time. Maybe some day the scientists will say that the unique plant ecotypes that are only found at Limahuli *should* be formally recognized by taxonomists.

Chipper told me that at that time, there was no formal garden or visitor program at Limahuli. In fact, some guy was living in the parking lot, and the beautiful old rock walls were covered with weeds. But spending time in the Upper Preserve inspired him to think more seriously about restoring the far more degraded Lower Preserve:

> Being up there was like being in another world. Everything was native; it was unreal, and it had a profound impact on me. I'd walk through the Lower Preserve and think how great it would be to restore it, even though I knew it would be a long-term process that I'd never live to see fulfilled. Doing the botanical surveys helped me realize how dominant the alien species were, especially the areas where the cattle had been [as was the custom in those days, the cattle ran wild throughout the Lower Preserve from 1875 until Chipper's grandmother fenced them out in 1967]. Yet there were still some really neat patches of natives left, and I started thinking that it wasn't totally hopeless.
>   Of course, at that time people weren't much into restoration; they were almost exclusively focused on preserving the best remaining places. If they thought about it at all, they believed that restoring degraded areas would just be a waste of time and money. But then my grandmother gifted the front thirteen acres to the National Tropical Botanical Garden, and one day each week this hippie guy and I would go to work—just the two of us against the alien forest! He couldn't believe what we were doing—to him cutting down all those plants was sacrilegious, like we were killing Mother Nature.

Eventually, Chipper went back to school and graduated from the University of Hawai'i at Mānoa in 1983 with a degree in tropical horticulture. (Although he hadn't done well in high school, by this time he was so motivated that he graduated with a 4.0 grade point average and represented the whole College of Tropical Agriculture at graduation.) He also became increasingly interested in Hawaiian history and the native culture and language, especially after he met Haleakahauoli, who was enrolled in the UH Hawaiian Studies program (Chipper's family is descended from the first white missionaries on Kaua'i, but he is also part Native Hawaiian). After graduation, both of them started working with the local Native Hawaiian community on Kaua'i's north shore and learned more about how plants were used and perceived within that culture.

When Chipper was finally ready to implement his larger vision for Limahuli, he suddenly realized that the whole valley actually lay within a highly restricted Conservation Zone, and ironically, everything that had already been done in the name of conservation was probably illegal. To proceed with his plans, he had to apply for a permit, which in turn required rezoning the entire region. "I was still highly naive about bureaucracy," he laughed. "I didn't even know anything about permits until I went to build a garage! So I wrote my first Master Plan for Limahuli and request for rezoning on a typewriter in 1986, and of course they rejected our application."

I knew from conversations with other people just how frustrating and convoluted that initial rejection had been, and that many people at that time were sure that the State Department of Land and Natural Resources would never rezone the area and grant them their permit. Yet in his typical indomitable, lemonade-out-of-lemons fashion, Chipper just went back to work and redoubled his efforts. Chipper admitted,

> It was a huge pain, but that rejection forced me to really think through the details of my vision and fine-tune my plans, and because of that, in the end we came out with something that was much better than my original application. I learned that you've got to be willing to let people criticize something, and that a lot of times you can really

improve what you want to do by listening to and addressing their criticism.

While that process was underway, my grandmother died and left the whole valley to Hauoli and me. We both felt that that land wasn't so much a gift as a responsibility, and it took a couple extra years to incorporate the entire thousand acres into our Master Plan, which ended up being over a hundred pages long and costing tens of thousands of dollars to complete. But after seven long years, the State gave us our permit at the end of 1993 after I *finally* managed to convince them that preservation isn't enough, and that without active management and restoration, our special ecological areas will just continue to degrade.

Looking back, Chipper felt that on the whole, things at Limahuli had generally progressed along the lines he had envisioned in that first Master Plan. Perhaps the biggest deviation from his initial plans has been within the Upper Preserve, which he had originally proposed to leave alone because of its pristine state at that time. However, after Hurricane Iniki knocked down most of the native canopy trees in the lower half of the Upper Preserve and opened the door to the establishment and spread of noxious alien weeds, he knew they would have to actively manage and restore that area as well.

There was so much I wanted to do in the Lower Preserve once we finally got our permit, but at the same time, we were also mandated to become financially self-sufficient, so I had to start by focusing on the Visitor Program to generate enough money to keep the garden afloat. Then, after that was up and running, I got that $15,000 grant from the Forest Service for you to do that research in the Lower Preserve into what methodologies might be most effective for restoring that mesic forest environment, so I held back to see how your experiments would turn out.

Matt, the Sierra Club volunteers, and I slung on our daypacks and headed up toward the outplanting area within the Lower Preserve's mesic forest. To get there, we first had to walk along the beginning section of the public trail that leads visitors on a self-guided, three-quarter-mile walking tour around its more formal "living classroom" section. I had taken this tour many times before, and I always learned something new and interesting from the descriptive signage along the path and from reading the accompanying information in their fifty-page tour booklet.

Matt led us off the public path and onto a narrow, unmarked trail that paralleled the stream and took us deeper into the forested valley. When we had gone around a sharp bend that hid us from the public, he stopped to explain that in the old days, the Hawaiians would ask permission to enter an area that they weren't from, just as we would knock on the door of a stranger's house. Because this valley was now unoccupied, he was going to offer a chant to the resident forest spirits and wait for some indication that it was OK to proceed. We must have collectively looked at this thirty-one-year-old white guy from New Jersey with some skepticism, because he then patiently explained that the local Hawaiians had assured him that what matters is not one's race, but rather one's intentions and heart.

He closed his eyes and earnestly sang a beautiful song in Hawaiian that was, at least to my untrained ears, entirely convincing. When he finished, we all stood motionless and listened to the strong wind rustling through the trees. Due to the valley's high walls, the steady northeast trade winds tend to spill over the tops of the adjacent mountains and swirl around forcefully within much of the Lower Preserve. *Lima* literally means "hand" and *huli* to "turn over or search." According to the authors of the oral history report I had read on the previous night, "Limahuli" is thus not only the name of this valley and its major stream, but also the "named wind" that ricochets around this place, turning and tumbling its vegetation like a "probing hand."

Apparently satisfied that we could proceed, Matt picked up his pack, and we followed him up the steep, rocky trail that led to the outplanting area.

Five years earlier, when a colleague and I first flew over from the Big Island to plan our restoration experiment in the spring of 2000, virtually the entire overstory of this area had been dominated by alien trees such as guava (*Psidium guajava*), Christmas berry (*Schinus terebinthifolius*), Java plum (*Syzigium cumini*), and octopus or umbrella tree (*Schefflera actinophylla*). This latter species is a common, well-behaved houseplant throughout the temperate zone, but it metamorphoses in Hawai'i into an aggressive, towering feral tree. The understory at that time had been blanketed by 'awapuhi ginger (*Zingiber zerumbet*, a nonnative species brought to Hawai'i by the Polynesians, who used it for shampoo, medicine, and to scent their clothing; several commercial shampoos today are made out of this plant) and some exotic grasses.

We had begun our fieldwork by flagging out sixteen plots in a homogenous section of that weedy forest. After debating and ultimately rejecting several more complex experimental designs, we had decided to simply assign four of these plots to a complete overstory and understory removal treatment, four to just overstory removal, four to just understory removal, and four to serve as controls (no vegetation removed). Within each of those sixteen plots, we marked out a series of smaller subplots for monitoring and supplemental treatments such as additions of native seeds and seedlings.

We kept this experiment relatively simple because we wanted it to be easy to interpret and provide unambiguous, practically valuable guidance to Chipper and his staff as they scaled up their restoration efforts in the Lower Preserve. We hoped our results could help them answer a series of basic yet critically important questions such as these: How much of the alien overstory and understory should be removed at one time? What kinds of natural recruitment of native and alien plants will occur within each of the different vegetation removal combinations? Which native species should we plant in there, which can be established by direct seeding, and what is the best microenvironment for each of them? Which weeds should we go after aggressively, and which can we safely ignore?

On a subsequent trip, I worked out the myriad logistical details of implementing this experiment with another Limahuli restoration staff member named David Bender. I had also managed to shake loose some internal Forest Service money to give to Limahuli to help defray their considerable in-kind costs associated with this project. In fact, Dave estimated that it would take about two weeks for three guys working full time just to perform our initial overstory and understory removal treatments.

In one sense, that research had gone well. Dave and his crew did a great job setting up and launching the experiment, and I always looked forward to flying over to see what was happening and help decipher our results on both an academic and practical level. Some of what we saw was quite encouraging, particularly the unexpected explosion of natural recruitment within a few of our treatment combinations by two "weedy" native understory species unique to Kauaʻi (the fast-growing shrubs *Bidens forbesii* and *Pipturus kauaiensis*). However, as was the case in virtually all of my other "straightforward experiments," the interpretation of this one turned out to be deceptively complex and inconsistent. In a nutshell, we ultimately found that the different native and alien species sometimes responded to the different treatment combinations differently.

Ironically, the strategically important area taken up by these sixteen experimental plots eventually became an obstacle to Limahuli's ongoing restoration program. Chipper and his staff were understandably eager to scale up their weed control and native outplanting efforts in that general area, and having to go around our plots and leave their often considerable populations of weeds alone became increasingly burdensome. Moreover, as Dave and his crew grew ever busier trying to stay on top of their numerous other "real" restoration projects, the time they could devote to that research steadily diminished. Thus even though we had originally intended this to be a long-term research project, we finally decided that it made more sense to let the field crew clear, weed, and restore that entire experimental area.

I asked Dave in 2005 whether he thought our experiment had been practically relevant to the on-the-ground restoration of the Lower Preserve. "I think that research helped us confirm some of our intuitive ideas about how to proceed in there," he reflected. "We probably could have learned all that with a less formal trial-and-error approach, but then we wouldn't have had that extra confidence that comes from doing things more scientifically."

However, when I asked him whether he had ever been able to extract any practical value from our more subtle, complex results that could not have been gleaned from a more informal experiment, he shook his head. "Not that I can think of. In doing that kind of work, on that kind of scale, we mostly had to deal with site-specific stuff, and the logistical complexities of getting people and equipment and plants in there. There just wasn't time or room for thinking about or doing much with the kind of complex ecological interactions that formally trained scientists typically focus on in the field."

When I asked Matt to reflect on his experiences restoring the Lower Preserve, I got a similar response. "We've mostly just learned by doing," he said matter of factly. He went on,

> We try plausible things, like different herbicides and different ways of ripping trees out of the ground, then watch what happens and adjust our practices accordingly. When I first started working here, the south shore gardeners would tell me things like, "Oh, it's critical to give that plant special fertilizer; you have to plant this one lying down and berm up the soil around it." All that may be true on the south shore, but it wasn't true here. For example, if I planted anything too shallowly, the winds up here would blow them right down. I ended up just planting those "special plants" like everything else, and maybe because of our fertile soil and abundant rains, every one of them has thrived.

When I mentioned some of the specific, more complex things we had discovered in that restoration experiment and asked whether those results and that kind of more formal science in general had been relevant to their on-the-ground efforts, Matt's response largely mirrored Dave's:

> Scientific knowledge is obviously important, and maybe down the road some day, when our program is further developed, we'll be able to do and utilize more of it, but right now it's really just practical experience. We do very little monitoring or data collection, unless we have to provide some formal documentation to a granting agency, because we're always just scrambling to get the work done. Sure, we've found things like some natives need full sun, and having some

canopy shade can sometimes slow the understory weeds down, help out some of the natives, and maybe let us get by with less watering. But what we actually do depends more on what we find in each specific area. For example, when there are few or no natives, sometimes it's easier to just come in and clear out everything. And we've found that in some places where there's fertile soil and a good native seed bank or adjacent population of native plants, if we expose some bare soil and stay on top of things with our follow-up weeding, the naturally recruiting, pioneering natives will eventually dominate the understory and create good habitat for us to come in later and outplant the things that would never establish on their own. Then again, some other seemingly similar areas have turned out to have very different dynamics, perhaps because the dominant aliens were different, and there were different natives in the understory, so in those instances we had to use a totally different plan of attack.

We also have to do things differently when there are still some residual natives present. Sometimes we'll try to slowly thin the aliens down around them, but that too depends on what the specific natives and weeds are. Some of those alien trees are seventy-five feet tall and have huge, spreading branches! We know we're eventually going to have to come back and take those trees down, and no matter how careful we are, we're almost certainly going to hit some of the surrounding natives in the process. In other places, we've also learned the hard way that we need to leave at least a few alien trees to serve as wind breaks.

I asked Matt how they decided what to plant after they had implemented their initial weeding and site preparation treatments. He explained that they usually start by simply looking at which if any natives are already there, then try to collect, propagate, and outplant them back into the field if and when they can. They also ask experienced field botanists what they think could and should grow in their different sites and rely on their own observations of what has and has not worked previously.

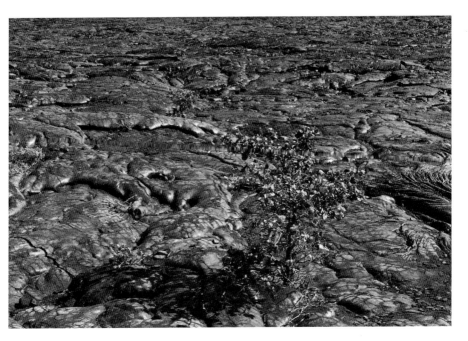

Plate 1. 'Ōhi'a (*Metrosideros polymorpha*) colonizing a barren lava flow. *Jack Jeffrey*

Plate 2. Native vegetation on lava with Mauna Kea in background. *Jack Jeffrey*

Plate 3. Gorse infestation on eastern slope of Mauna Kea. *Jack Jeffrey*

Plate 4. *Nēnē* goose (*Branta sandvicensis*). *Jack Jeffrey*

Plate 5. ʻIʻiwi (*Vestiaria coccinea*) on the endangered lobelia *Clermontia lindseyana*. Jack Jeffrey

Plate 6. Volunteer with koa seedlings for corridor. *Jack Jeffrey*

Plate 7. 'Akiapōlā'au (*Hemignathus munroi*) in koa corridor. *Jack Jeffrey*

Plate 8. 'Amakihi (*Hemignathus virensin*) in koa corridor. *Jack Jeffrey*

Plate 9. Lava consumes remnant native forest. *National Park Service*

Plate 10. Park Service staff clearing an invasive stand of alien ginger. *National Park Service*

Plate 11. Aerial view of Auwahi One in foreground, Auwahi Two in background, 2011. *Art Meideros*

Plate 12. Limahuli's Lower Valley viewed from above. *Dave Bender*

Plate 13. A section of Limahuli's Upper Preserve fence. *National Tropical Botanical Garden*

When I asked him how they determined the best methods for propagating and outplanting their native plants, Matt threw up his hands.

The truth is that all those methodological issues are largely driven by logistical issues. We're always behind, racing to get the plants out the door because the nursery is bursting at the seams and we need to grow a bunch of new plants for another project, and the plants are starting to get root-bound or too heavy to carry and we can't get the machinery into our field site because the trail is too steep, or it crosses an old stone terraced wall or an archaeologically important feature. So we end up just doing the best that we can with our very limited time and resources.

Once the plants are in the field, we always tell the volunteers to use their best judgment, common sense, and intuition in terms of microsite placement, spacing, species mixture, et cetera. We never know how many of our outplants will survive, and we see very different patterns in nature, so who can really say what the "right" way to do it is? Sometimes we'll get lots of herbaceous natives popping up within the same holes we dug for our trees; sometimes they really seem to help them, other times they don't. But one thing we've consistently found that doesn't work for us is trying to direct-seed our plants in the field. Even when they germinate well, we lose them in the weeds unless we're constantly on top of them, which never happens because we just don't have the time or staff to do that.

As the Sierra Club volunteers and I walked into the first outplanting site, it was easy to appreciate why they had moved away from direct seeding because the entire area was a chaotic jungle of interspersed native and alien trees and shrubs and herbs. I couldn't imagine how anyone could possibly even find, let alone weed, a bunch of wimpy native seedlings within such a mess.

Matt explained that our mission was to weed around the smaller outplants so that his crew could spot and hopefully not hit them with their weed whackers the next time they swept through this site. He demonstrated how to do this by walking over to a little native sapling and deftly ripping out all the weeds around it with his hands. As he was working, I noted the confused expressions on a few of the volunteers clustered around him.

"But how do we find the plants to weed?" a middle-aged man finally stammered.

"Well, they're all over the place here," Matt replied, apparently not realizing how invisible they were to the inexperienced members of the group. He walked around and patiently pointed out a bunch of different outplants scattered throughout the area. "Just spread out, go slow, do the best you can, and of course let me know if you have any questions or need help finding the plants. Whatever you can do will be a big help to us."

I was too anxious to see how the area as a whole had developed to stay in one place and weed, so I wandered off by myself. The forest had changed so much in the years since my last visit that at first I couldn't even find the section where our restoration experiment had been. When I finally got my bearings, I walked over and stood beneath a remnant old *lama* (*Diospyros sandwicensis,* a member of the ebony family that produces edible persimmons and very hard wood that the Hawaiians fashioned into rafters and traps for deep ocean fish; they also pulverized the wood and mixed it with other materials to make compresses for the treatment of skin sores). That tree had been in the center of one of our research plots; although it had been the largest of the few surviving native canopy trees in that section of the Lower Preserve, we hadn't even noticed it at first because it had been engulfed by a tangled web of alien vegetation. But now, all that remained of those giant alien trees was their stumps, and even these were starting to disappear beneath a dense thicket of volunteer native shrubs and an impressive array of outplanted native trees.

I sat down beneath that *lama* tree and looked around. It slowly began to dawn on me that I was actually sitting in the middle of a regenerating lowland mesic forest dominated by native Hawaiian species! From my perch I could see several thriving outplanted endangered species, including a large block of *ālula* (*Brighamia insignis*). This succulent member of the bellflower family, often described as a "cabbage on a baseball bat," is one of the most unusual species within the entire Hawaiian flora (and that's saying something!). It was discovered over twenty-five years ago by two of the National Tropical Botanical Garden's plant collectors while they were rappelling down some remote cliffs on Kaua'i. Since then they have regularly gone back to monitor the wild populations, fill in for its presumably extinct pollinators, and collect seeds for propagation. Fortunately, *alula* has turned out to be relatively easy to propagate and outplant (its distribution was almost certainly far more widespread before ungulates and other alien species apparently extirpated it from everywhere except those few unreachable cliffs), and today the considerable number of outplanted specimens at Limahuli and elsewhere far exceeds the number remaining in the wild.

As I observed the thick clusters of native species growing up and over the relatively tamed understory weeds, I remembered something Phyllis Somers had told me years earlier, before any of this restoration work had begun. At that time she had had a vivid dream in which she was floating above the Lower Preserve, and when she looked down, she saw a completely restored, healthy, native forest. When she told Chipper about it the next day, he immediately said, "Keep dreaming, because we're going to do it!"

I walked over and rejoined the group to see how they were faring and contribute some of my own labor. Most of the volunteers had enthusiastically dived into the project and were finding and rescuing a lot of outplants. Some were putting forth an honest effort but appeared to be more interested in being in the forest than weeding it. A few looked as if they wanted to work but didn't know what to do or were afraid of accidentally killing some rare native species. I watched an elderly woman tenderly remove each and every damaged leaf she could reach from a small cluster of alien guava saplings. She told me that she didn't know any of the species in this forest, but she loved plants and had decided to try and help them all.

About an hour later, Matt thanked us for our work and said it was time to hike to the waterfall at the back of the Lower Preserve. Along the way, he excitedly pointed out a few of the remnant native plants we encountered in the sea of aliens, but people's level of interest in and appreciation of these generally nonshowy species appeared to be proportional to their prior botanical knowledge and general ecological education. However, everyone seemed interested in Matt's stories about the Hawaiians' relationships with and use of their flora. For instance, when he showed us a clump of nondescript, diminutive trees and identified them as *pāpala kēpau* (*Pisonia wagneriana,* a member of the four-o'clock family found only on Kaua'i), only a few volunteers bothered to walk over and take a closer look. But then he told us the story of the ancient bird catchers, who spread the sticky sap from this plant's fruits onto a branch, attached a flower as bait, and then pushed the branch up into a tree. When a bird perched on the branch, it apparently would get stuck long enough for the bird catchers to climb up and harvest a few colorful feathers before letting the bird fly away unharmed. When he finished his story, everyone immediately came over to examine this plant and test the stickiness of its sap.

Unlike the members of our group, who all appeared to uncritically accept this story, some of the people I knew in the Hawaiian research and conservation community would privately smirk whenever this "fable" was repeated. "Yeah, right," someone might say, "they made all those cloaks, containing all those hundreds of thousands of feathers, by climbing up and down the trees to get one or two feathers at a time? And they didn't kill any of those flightless birds either!"

One day I asked Chipper what his take on such stories was. "Well," he said thoughtfully, "the truth is probably somewhere in the middle. There may have been some times and places when the bird catching was abused, and other times and places when it wasn't. So much of our knowledge about the Hawaiian culture comes from the period of time after Captain Cook arrived, and yet we know that Cook's arrival changed everything, so I think it's important to keep an open mind."

I was still thinking about Chipper's words as Matt led us into a dense, fifteen-acre bamboo forest. Once again I found myself trying to imagine what the world of the ancient Hawaiians who lived here was really like. (Archaeologists believe they may have lived at Limahuli for well over a thousand years; why they left such a hospitable place remains a mystery.) After we had walked deep into this grove, we stopped and listened to the eerie sounds of the thick bamboo stems clinking against each other as the wind caught their leafy upper branches. Because the understory was so clear and devoid of other plants, it was easy to make out the old stone taro terraces running across the valley.

Even though I knew this bamboo was a nonnative invasive species that had been brought to Hawai'i by the Polynesians, who used it primarily for construction and tools, it was undeniably pleasant and enchanting to be in that forest. In fact, at the end of the day, several of the volunteers said that being in that bamboo was by far the highlight of their day at Limahuli. I wondered how the Hawaiians had felt about it, and whether we should just leave it be now or clear and replace it with native species.

When we reached the waterfall, Matt wisely gave everyone time to "ooh" and "ahh" and snap a million pictures that I knew from previous experience would fail to capture the grandeur of the place. After we'd finally had our fill, we walked back down the valley and reassembled on a little island in a wide, cleared-out section of the stream. Matt told us that this stream was one of the last in the state that had both high water quality and all five species of native freshwater fish. These fish, called *'o'opu* by the Hawaiians, evolved from marine ancestors in the goby family. Adult *'o'opu* live in freshwater, but their fertilized eggs wash downstream, and the newly hatched larvae spend their first several months in the ocean. To complete the life cycle, these young fish must then find and ascend a suitable freshwater stream.

Four of the five *'o'opu* species have partially fused pelvic fins that form a suction-cup–like structure that helps them cling to rocks and hop their way up the beds of rapidly flowing streams. Two of these species, the clinging and red-tailed stream gobies (*'o'opu nōpili* and *'o'opu hi'u kole*, or *Sicyopterus stimpsoni* and *Lentipes concolor*, respectively), can even use this structure to ascend the vertical rock walls of waterfalls. However, Matt told us that researchers had found that the fish couldn't get past a section of the stream that was choked by a dense thicket of *hau*. Apparently the *'o'opu*'s suction cups were foiled by a thick layer of silt that the slow-moving water had deposited over the rocks in this section of the streambed.

Shortly after I first landed on Kaua'i in 1996, some colleagues and I had hiked to this waterfall with Chipper. Early in that hike, Chipper had pointed out several crystal-clear pools that he used to swim in with his grandmother when he was a boy. But later on, as we were slogging through the stagnant, muddy water that ran beneath a jungle-gym–like barricade of tangled *hau* branches that extended across the stream and onto its steep, rugged banks, Chipper had told us that someday he was going to do something about all those thickets. Thus I wasn't surprised to learn that years later, when he heard about what was happening to the *ʻoʻopu,* he immediately said, "Let's get a grant to get it out of there now!" On this day, Matt recalled, gesturing all around the island,

> There was a half acre of solid *hau* here, and it was a silty, mosquito-infested mess. At first we talked about airdropping some heavy machinery in to clear it out, but then we were told we couldn't do that because of the archaeology sites around here and because that machinery probably would have silted up and damaged the streambed even more. So we just got a big crew of volunteers and used brute force. We started with handsaws and kept knocking stuff down and pulling it out until the water finally started trickling through again. The whole time we were clearing out the *hau,* the stream would shift one way, then flow the other way, over and over. Then we got some big floods that washed away the stumps and pushed out the silt, and the stream really started flowing hard again. Now it's back in its old channel, and within less than a year we found that all five *ʻoʻopu* species were making it back up to the waterfall.

I had heard a few incredible stories from the people who worked on that project and had seen some before and after photos, but I had still always had trouble believing they had really done it until I finally saw the stream's clear bubbling water and wide-open banks that day with my own eyes. I remembered how sore and exhausted I was by the end of that first hike with Chipper through the *hau* and how intractable I had thought that situation would always be. I looked up from the stream and over at the steep, weed-infested slopes and vertical cliffs of the mountains lining both sides of the valley and wondered whether if, in the end, nothing really is impossible.

"It is amazing what you can accomplish when you have the will and the people to do what it takes to get the job done," Chipper told me. "It's really gratifying for me to have had this vision for so long and get to live to see it implemented. The hardest part was just getting that vision off the ground, but once we did, things started rolling. Now the VIPs come out here and are blown away! There's nothing like success to create more success."

One early and important measure of their success came in 1997, when the American Horticultural Society named Limahuli Garden and Preserve "The Nation's Best Natural Botanical Garden" because its "research, teaching, and educational programs over the past ten years have demonstrated the best sound environmental practices of water, soil, and rare native plant conservation in an overall garden setting." Phyllis recalled,

> That award, along with some of the great press we got, was a huge turning point. It really helped us build the Visitor Program, which in turn gave us the financial ability to start the serious restoration work. All the grants are great, but they come and go, and they always want you to do something new. But every time we clear and plant another acre, we generate that much more ongoing maintenance that can't be neglected just because someone gave us money to start something else.

As was the case in so many of my conversations with other people in the environmental community throughout the Hawaiian Islands, the subject of money came up early and often when I spoke with Limahuli's staff about their future. Many stressed that the garden just did not have enough money to do all of the things they felt were critically important to do. Several also told me how hard it was to make ends meet on their modest salaries. Especially in the beginning, as the garden was developing, these kinds of financial problems were creating challenges for Limahuli's on-the-ground work as well as its institutional cohesion. Summing up the concerns of many of his colleagues during the first few years of that program, Matt told me in 2005,

Right now, our whole restoration program is floating economically. Like every other conservation organization, we never get enough money for the ongoing maintenance of our existing projects. It's frustrating, and scary, because if we don't stay on top of the areas we've already cleared, we could get some weeds in there that would be worse than what was there before we started.

Limahuli is an amazing place; we're like a family, and I love working here! We get great people, train them up, and create highly skilled, knowledgeable, motivated, and efficient work crews—it's a joy to be a part of all this. At the same time, however, it's awful when the next grant doesn't come through in time and we have to lay someone off, especially when that person is someone who was born and raised here, has a good head on his shoulders, and is working his butt off. People begin to realize that there's no security in this kind of work. That's why it'd be so great if we could get some permanent money for the program, so we could hire a large, permanent crew. But the reality is that right now there are only three of us working in the Lower Preserve, and none of us are on hard money. If someone gets sick, or hurt—and that always happens eventually, because this is hard, physical, sometimes dangerous work—then we're down to two, or one, which isn't safe. For the first time in my life, I'm seeing a chiropractor and a masseuse, because by the end of the day, I'm really hurting.

Many of Limahuli's staff members also stressed how expensive living on the island had become. Even in a state known for its exorbitant "Price of Paradise" cost of living, Kaua'i's north shore is notoriously pricey. Yet paradoxically, there are few decent jobs in this area, and even the better ones seldom pay enough for outsiders (or island residents that don't already have a free or relatively cheap place to stay) to pay their own room and board, let alone buy a house and raise a family.

One solution to this problem is for places such as Limahuli to hire local people. In fact, even if there had been a surplus of highly qualified and motivated outsiders who would have been able and willing to work at Limahuli for what the garden could pay them, Chipper would still have tried to hire as many locals as possible. "From the beginning, it was always important to him that the garden fully embody the *ahupuaʻa* concept," Phyllis explained, "and that includes getting local people involved and giving them key roles to play, because historically, part of the genius of *ahupuaʻa* was that they were controlled by the people who personally benefited or suffered depending on how well they managed their resources." All of the staff I spoke with strongly supported this philosophy, and several discussed how gratifying it had been to work with and get to know some of the local people who had deep personal histories with and strong emotional and cultural connections to Limahuli (some have genealogical connections going back over a thousand years).

Yet some told me about the tensions involved in working with locals—and each other. "My job as the assistant director here can be very challenging; I wind up mostly dealing with personnel issues," David Godale confessed. David, who is Chipper's first cousin, also grew up spending time with and working for their grandmother at Limahuli. When Chipper became the director of the entire National Tropical Botanical Garden in 2003 and moved to the south shore, he asked David to step in and serve as the interim director at Limahuli. David continued,

> Some of the locals on our staff work here mainly because it's a relatively good job and it's close to their home. Some are outstanding, of course, but others may not have the education or skills to do the job we hired them to do, and that can lead to some tough situations. It would be great if we could have one large, harmonious group all working together, but the reality is that the different interests and abilities of our staff as a whole can create tensions within and among the work crews.

These kinds of personnel challenges are ubiquitous throughout Hawaiʻi's environmental community and are often exacerbated by the different educational backgrounds and employment trajectories of "outsiders" versus "locals." For instance, Dave Bender told me that he had to deal with resentment and insubordination from some of his former less-educated, local coworkers after he was promoted to a supervisory position and started spending more time writing grants and managing computerized databases. "The guys on the crew didn't respond to my directions once I was in the office," he lamented. "They didn't like my not being out there with them doing the physical work, and they didn't understand or respect the value of the office work I had to do."

Another potential solution to the perpetual money and personnel problems is to create an ambitious volunteer program. Like other highly successful Hawaiian conservation-oriented organizations, Limahuli has been able to make extensive and effective use of their volunteers. "They are absolutely critical to our restoration program," Matt said. "It takes a lot of time and energy and resources to train and manage them in the field, but the payback is well worth the trouble. We'll probably get more done in the seven weeks we'll be working with the volunteers this summer than the three of us on the restoration crew can do in the entire year."

As great as they can be, however, volunteers have their own share of challenges. First, it can be difficult or impossible to know ahead of time how willing and able they will be to do the often hard and tedious work that needs to be done. "We've had many absolutely amazing volunteer groups here," Nancy Merrill told me. "They were a joy to work with from the moment they arrived, and we hated to see them go at the end. But we've also had a few groups that were less capable and motivated; you just never know how it's going to go until they show up."

Second, even the best volunteers require considerable amounts of training, supervision, and logistical support. Consequently, effectively recruiting and utilizing them requires a substantial amount of time from staff members who typically are already overwhelmed by all their other responsibilities.

Finally, because they are "only" volunteers, it can be risky to depend on them. Much like money, the availability of competent volunteers ebbs and flows over time, especially for a remote garden on a small island in the middle of the ocean. In addition, some essential tasks don't lend themselves to large groups of inexperienced volunteers, or they must be done during times when few volunteers are available. "Our volunteers have been great," Chipper concluded, "and we certainly couldn't have accomplished as much as we have without them. But we'll always need our core of highly trained and motivated professional staff to ensure that things are done right, keep our various programs on track, and provide the overall institutional continuity."

Given the perennial challenges of securing sufficient funding, recruiting, training, and retaining good staff and volunteers, not to mention figuring out what to do and how to do it on the ground, what is a poor restoration manager to do? What is the right balance between detailed plans and rigid, standardized protocols versus a more liberal, go-with-the-flow philosophy? In the grand scheme of things, where does formal scientific research and knowledge fit it? Chipper acknowledged that

> It's tough to find the right mix, and there's often tension among the different components. When you have a very limited pool of money and resources to draw from, should you launch a scientific research project or break out the chainsaws? Looking back, I'm glad we started with the science. But as we started getting more grants and more money from our Visitor Program, I didn't feel a need to keep doing the science, because restoration is really more of an art than a science, especially for the guys on the ground.
>
> For instance, when we started restoring some of the ancient rock work in the Lower Preserve, we worked out an arrangement in which the Native Hawaiian stonemasons and the archeologists would have an equal voice, because as artisans, the stonemasons could see things in the alignments of the stones that the archeologists couldn't. Similarly, as the restoration program progressed, I realized that we needed to bring the practitioners into the visioning and planning processes, because so much of it is unpredictable, and we need to be able to adapt to what's happening on the ground. I had some pretty specific, detailed restoration ideas and methodologies in my original Master Plan, and we've since

written many grants in which we propose to do things in a certain way. But one of the real challenges for me has been that when we started doing the actual work, the guys on the ground would often find that it couldn't really be done that way, or another way would be much better. So I've always tried not to box myself in by being too much of a purist. Having said all that, however, I have also always maintained a certain level of overall discipline and accountability. For example, I have been careful to make sure we are only using appropriate plant species with appropriate genetic backgrounds in all of our restoration plantings.

When I asked Matt to reflect back on the extent to which he had been able to implement the Lower Preserve restoration program according to any preconceived plans and timetables, he just laughed:

Understandably, people always want schedules, what you're going to do that week and how you're going to do it, et cetera. I tried to do that a few times here, but then I stopped, because it was a waste of time. I started to feel like I was back in the federal system, where we had to lay everything out ahead of time, even though there was almost always a night-and-day difference between what was on paper and what actually happened. Fortunately, Chipper gives us the freedom to get things done the way we think is best. So at Limahuli, I did away with the schedule; we played it week by week—or more often, day by day. I might think that we were going to, say, herbicide an outplanting site that desperately needed it, but then it'd be too windy or rainy, so all of a sudden we're hand-weeding in another area. But we end up being far more efficient this way; that's why lots of groups, including some of the government agencies themselves, prefer to fund us, because they know they'll get more bang for their bucks.

When I asked the Limahuli staff members what they felt their most important accomplishments have been so far, nearly everybody mentioned their education and outreach programs in general and their work with children in particular. David Godale told me,

> Almost every group that comes here gets engaged in a meaningful way. Different people get hooked by different things—the majesty of the place, the medicinal value of the plants, the conservation mission, the cultural connections. When the kids come here, they finally get the concept of *ahupua'a*. We've found it's best if we can get them working—touching and feeling things and getting dirty—and we always try to end with some planting. Plants have a lot to tell you, and many of the kids pick up on that, but you've got to get them close and still enough to listen. When they come back years later and see the effects of what they did, and especially how the plants they planted have grown, there's tremendous power in that experience—it's like a freight train!

"I visited a garden when I was twelve," Nancy recalled, "and it turned me into a lifelong lover of gardens and plants. I know we are doing that for some of the kids who come here, and maybe some of the adults too, although they can be harder to reach." While I personally could not imagine a more effective and enjoyable place to learn about Hawaiian ecology, conservation, and culture, I knew from my own observations that at least a few of their visitors did not appear to get much out of their time at Limahuli or appreciate the quality of their programs. As Nancy explained,

> You can tell what kind of experience they're going to have before they even go out. Some tourists are in such a hurry trying to tick off all of the things on their list, you can tackle them and they won't get it. Fortunately, they're the exception, because we tend not to attract those kinds of people in the first place. In fact, we could easily double the volume of people coming here, but that's not the kind of experience we want our visitors to have. Right now most people don't see anyone else on their tour, and that's how we'd like to keep it.

Phyllis added,

> Some of the men are dragged here by women who want to see some beautiful tropical flowers. Ironically, once we explain that this garden is not really about flowers, some of these women look disappointed, but then the men become more interested! That's the hardest message to get across—that a garden can be about something other than just flowers. Sometimes I'll point to the hillside across from the Visitor's Center and tell them that area used to be all poinsettias, but we replaced them with native plants that have their own unique beauty and value, and then some of them will start to understand what we're all about.

Phyllis and Nancy both stressed that their jobs have become progressively easier over the years because the garden has gotten better and because their visitors are more informed than they used to be. Phyllis observed,

> There is also a sacredness in the land at Limahuli. This sacredness was passed on to Chipper from his grandmother, and he's passed it on to those around him. And he does things extremely well—he's very patient, there are no loose ends here—and most of our visitors pick up on all that. Every day I'll see a few visual transformations; people say they somehow feel cleansed—and spiritual. Then they'll often say they had no idea about any of the stuff we're trying to teach here, and they'll ask what they can do to help. You know, all the professionals would say we've done everything wrong—the tour is too long, the self-guided booklet too detailed. But when we see how deep and meaningful so many people's experiences are, when we see that glow in their faces when they come back and hear their stories about how this was the best thing they've ever done in Hawai'i, we know it's all worth it.

After being away for several years, in the spring of 2012, I asked

Kawika Winter, who became Limahuli's new director in 2005, to update me on how things at the garden had developed and evolved during his tenure. He said immediately,

> We've grown by leaps and bounds! For one thing, the steadily increasing revenue generated by our Visitor Program has enabled us to substantially improve and expand several aspects of our program. For example, in 2006 we created a new interpretive section called the "Plantation-Era Garden," which we use to display and interpret the iconic showy flowers and edible fruits that were brought to Hawai'i during that period. In addition to being a fun and effective way to educate our visitors about the species that are *not* native to Hawai'i, we no longer hear any complaints from the people who came to see some pretty flowers!
>
> To better showcase our conservation work, in 2007 we transformed a half-acre section of the publicly accessible portion of the garden into a demonstration of our restoration work in the back of the valley. That year we also received the prestigious Koa Award from the Hawai'i Tourism Authority for our "long-term and exemplary commitment to perpetuating and preserving Hawai'i's host culture." And then in 2010, we got a grant from the Office of Hawaiian Affairs to do a restoration project to provide native plant materials to Hawaiian practitioners of culturally important activities such as lei making, herbal medicine, and wood carving. That grant enabled us to double the size of our original half-acre demonstration project.

Kawika told me that shortly before he came on board, Chipper placed Limahuli's conservation program under the direction of David Burney, a senior research scientist he had hired in 2004 to be the director of conservation for the National Tropical Botanical Garden (NTBG) as a whole. Part of the motivation behind this change was to improve the amount and dependability of funding for conservation on an institutional level. Sure enough, under Burney's leadership, the conservation department has since been able to acquire enough money to largely solve the financial and staffing problems they had been struggling with during my previous visit. For example, since 2005 they have secured a string of relatively large, long-term federal contracts for understory restoration projects in the Lower Preserve that have greatly expanded their work in that area and improved their overall institutional stability. In fact, the NTBG has not had to lay off any of their Limahuli-based conservation staff since Kawika became this garden's director.

Motivated in part by these new federal contracts, the revamped Limahuli restoration crew began shifting their overall strategy in the Lower Preserve. Although it required a tremendous amount of time and effort, they found that clearing the forest floor of invasive alien weeds enabled them to establish a broad diversity and impressive abundance of native understory plants, including several endangered species that were teetering on the edge of extinction. The success of these efforts has also attracted new researchers to Limahuli, which in turn has led to important new discoveries. For example, in late 2010 two University of Hawai'i entomologists discovered eight new species of native moths in the Lower Preserve, and most of these moths were found within areas that had been subjected to their more intensive understory restoration efforts.

In 2006, a team from the Kaua'i Endangered Seabird Recovery Project discovered that two rare, ground-nesting native seabirds, the critically endangered Hawaiian petrel and the threatened Newell's shearwater, were nesting in Limahuli's Upper Preserve. This marked the first known nesting site for the petrel on Kaua'i and one of only a handful for the shearwater. These discoveries helped Limahuli ultimately secure the roughly $750,000 they needed to construct their long-dreamed-of ungulate fence around the entire perimeter of the four-hundred-acre Upper Preserve (see Plate 13). They have also been able to secure a substantial amount of multiyear mitigation money from their local electric utility to help offset the negative impact of that company's power lines and lights on these two seabird species.

Consequently, while trying as best they can to preserve and restore a suite of other endangered animal and plant species that reside in this hanging valley, most of the efforts of Limahuli's Upper Preserve crew are now devoted toward bird-related activities such as

Landscape view of Upper Preserve fence.
National Tropical Botanical Garden

Restored native forest understory in the Lower Preserve.
National Tropical Botanical Garden

exotic predator control and habitat improvement. However, the bird money has enabled them to install housing and other much-needed infrastructure in the Upper Preserve, which in turn has catalyzed a series of productive collaborations with sea- and forest-bird specialists, entomologists, snail biologists, and other researchers interested in this previously inaccessible and largely unexplored native ecosystem.

Finally, Limahuli's staff has been involved in various marine conservation efforts as part of their larger "mountains-to-the-sea" *ahupuaʻa* program. Their long-standing collaborative work to maintain the health of their adjacent coral reef ecosystem while supporting the consumptive needs of local people through culturally rooted management activities resulted in the governor of Hawai'i signing a law in 2006 officially designating their *ahupuaʻa* as a "Community-Based Subsistence Fishing Area" (one of only two such designations in the entire state). As Kawika concluded,

> We're proud of all the things we've accomplished since your last visit and very excited about all the things we're doing now. Solving the major challenges that had been holding us back—the staffing and funding issues, institutional structural problems, on-the-ground methodological/technical shortcomings—has enabled us to make tremendous progress! Of course, there's never any shortage of new challenges. For instance, this year, we're finally doing a feasibility study for the development of a guided tour through our restoration areas and up to the waterfall. That was actually part of Chipper's original long-range plan, and now there's so much more to see along the way than there was back then. But as we consider this and countless other potential new projects, we often have to struggle to find the right balance between meeting our most important long-term, overarching goals, such as reconnecting our surrounding community to the forest and reawakening the Native Hawaiian traditions of natural resource management, and fulfilling the more concrete, shorter-term deliverables typically required by our funding agencies. Similarly, as we've grown and expanded in so many different directions, it has become more difficult

to integrate and align the often distinct methodologies, bureaucracies, and ideologies of the different component pieces such as our science, restoration, education, culture, and visitor programs. Consequently, Chipper and I are now working on developing a long-overdue integrated vision and management plan for the biocultural resources of the entire Limahuli Garden and Preserve.

Chipper added these final comments:

I see Limahuli as a place of refuge. The odds against us are huge, the problems are immense, and our resources are limited. But at the same time, look at what we've already accomplished, and all the ongoing and upcoming projects that are poised to bear even more fruit! And we now have a highly trained, educated, and experienced staff in place, with access to techniques and technologies that hadn't even been invented when we first started doing this kind of work. I also believe that the future will bring ever-more knowledge and breakthrough tools and ideas that we can't even imagine today. In the meantime, I want to keep ramping up our work on the ground, building upon our many successes, and of course, keep bringing the kids to Limahuli and getting them engaged. That's what my grandmother did for me, and those experiences have had a profound effect on me ever since.

Part 3   **Herding Cats with Leaf Blowers**

9   MULTIPLE PERSPECTIVES

What makes people dedicate their lives to preserving and restoring Hawai'i's remaining native species and ecosystems? Despite their seemingly common causes and struggles, why is there so much contention and even animosity among some of the individuals within Hawai'i's larger environmental community?

To explore these kinds of topics, I conducted extensive interviews with a diverse group of people working in Hawai'i's environmentally related academic, conservation, education, regulatory, resource management, and scientific research communities. In addition to our typically far-ranging, general discussions, I asked each interviewee four overarching questions:

1. Why do you care about biodiversity in general and Hawai'i's native species and ecosystems in particular, and what motivates you to try to preserve and restore them?
2. What is your restoration philosophy?
3. What do you think are the most important things we

should do to preserve and restore Hawai'i's natural environment?
4. What role does science presently play in guiding ecological restoration in Hawai'i, and what if any changes would you like to see in the future?

Hawai'i's broad environmental community is comprised of many dedicated and passionate groups and individuals that sometimes splinter into factions that do not always respect and support one another. Indeed, I learned the hard way how quickly one can inadvertently burn key bridges that are difficult or impossible to repair. Consequently, in exchange for their candor, I promised my interviewees that I would edit their answers and obfuscate the details of their specific situations as necessary to preserve their anonymity. In some cases, I have also combined the responses of two or more interviewees when they were sufficiently similar and complementary.

Question 1: Why do you care about biodiversity in general and Hawai'i's native species and ecosystems in particular, and what motivates you to try to preserve and restore them?

While everyone's answer to these somewhat personal questions was unique, many of the people I spoke with (as is the case with environmentalists in general) said that their interest in and concern for biodiversity had grown out of their childhood love of nature and the outdoors. Interestingly, however, virtually all of the people who grew up in Hawai'i had little to no knowledge throughout their childhoods of basic conservation issues:

My friends and I really weren't very aware of what was what—we had no idea that the mango and banana trees we'd find in the rain forest weren't native to the islands. We didn't even have any awareness of the concept of native and alien species, let alone understand anything about Hawai'i's ecological and evolutionary uniqueness. But we spent all our time outside, and it was all just one big green and beautiful playground.

Over time, something inevitably sparked their curiosity and led them to learn more about the natural world they once took for granted. For instance, one person recalled that when he was a little boy he spent all his free time fishing, but as he grew up he noticed that he was catching more tilapia and fewer native species. Similarly, when he started hiking, he realized that the wildlife along the remote trails was different from what he saw in the more developed, accessible areas. For others, their initial attraction to conservation was motivated more by a personal connection:

My interest was more cultural than ecological; the Hawaiian species were the true natives; they were here before the first humans. Then I started learning the Hawaiian language and working with Native Hawaiian people and learned how the plants were used, what their meanings were, et cetera, and I just became engulfed in that whole perspective of having a cultural responsibility to love and care for the natural world.

Those who came to Hawai'i later in life frequently discussed how they fell in love with the islands and the culture and just wanted to help take care of them and give something back:

When I first came out here, I immediately felt more of a connection with the Hawaiian people than I ever did with my own—I kept peeling back my skin to see if maybe I was part Hawaiian deep down inside! I still love Hawai'i's culture and natural environment—they've given me a wonderful life. And I love to hike, and look at things closely, and learn more about the things I see in nature. I want to help protect the natural world so that others can have the same opportunities that I've had.

Of course, some people just looked at me as if I had asked why they cared about their friends and family. When pressed, three such individuals responded as follows:

It just seems like the right thing to be doing at this time of my life. It's my service, my mission, my duty to give back in the world and make it a better place.

I get tired of talking about how I feel about the Hawaiian environment. It's a personal thing that's no one else's business. But I have this connection, and I want others to have a connection, whether it's here in Hawai'i or elsewhere.

I can't explain it, I just do. I always have, and I always will. The value of nature speaks for itself, and there's nothing more to say about it.

However, some were willing and able to provide more detailed answers that typically included a rich mixture of values, emotions, utilitarian factors, intellectual interests, and spirituality:

Nobody is ever going to get rich working in conservation—you have to do it because it's where your heart lies, where your values are. I just have to keep trying—I can't give up. I want people to realize this is not just an issue for the tree huggers! I don't want to have to pay $5.00 for a gallon of desalinized water from the ocean or have to start dealing with fire ants and snakes.

I'm always thinking about the way things were out here before humans arrived—the wild and crazy paths evolution traveled—and I just want to preserve as much of those things that make this place unique, especially because the rest of the world is becoming so homogenized. So I guess my motivation is purely aesthetical—preserving biodiversity just makes me happy; it's biophilia.

I came to care because of the experiences I had with my grandfather when I was a kid. He took me into the woods and taught me how to hunt and fish, and he had a very strong sense of place and love of nature that had been passed down to him by his father and grandfather. He passed these things on to me, as well as an appreciation of the difference between taking care of the land and abusing it.

As a graduate student, searching for hours in the rain and mud for some bloody little thing and seeing one, or two, or more often, none of them led me to develop a special affinity for Hawai'i's most desperate cases. I just love the underdogs, the species destined for extinction. I watched the warm fuzzies get all the money and attention while everything else was going before our eyes, and I decided that someone's got to stand up for the rest.

Although I was raised as a Catholic, I've always seen God in nature and believed there was a higher power out there. I have a personal relationship with particular native species; being with them, and experiencing the natural balance and beauty of intact native ecosystems . . . it just gets me high.

Whenever I'd see the park guys after work, they'd be all smiling, laughing, having fun, and I thought, "Hey, I want to do that!" Growing up out here, we were only interested in the plants we could eat, but now my heart is in it; I really wish we could get rid of all these alien species. Because of them, much of Hawai'i now looks like hell. You shouldn't abuse one thing and let the other thing go; there should be a balance. It's good to share, like the people born here who have so much aloha. I think they were inspired by the native plants.

I just wanted to find a career path that wasn't destructive. I've always loved growing plants and putting them in the ground. When I discovered that people get paid to do that kind of work, that really got me motivated! I love my job; I get to hide out here in this park, where our whole reason for existence is to do this stuff, and stay out of all the controversies and confrontations that occur in the outside world.

In the future, people can look back and say, "Hey, this plant was extinct in the wild, but people took the effort to save and restore it." That's all the motivation I need to keep going. I'd like to think if I did anything of value in this life it will have been to resuscitate some fossils. All of them won't get through, but if we don't try, none of them will.

My mentor always said that "Research is creative play. If you're not having fun, you're not doing good work!" Although many will publicly say that we're doing God's work, most of us will admit after the second beer that we do what we do because it's fun and interesting.

I look at something like the space program and wonder why we're spending all that money on that stuff, but I know that others look at conservation and ask the same question. Because ethical or religious or emotional/spiritual reasons for saving native biodiversity are so subjective,

I prefer the utilitarian stuff. Nevertheless, I see us as the modern missionaries: we come into a place and play God, even though we obviously don't have all of God's knowledge—just enough to be dangerous! But what else can we do? If we care, we really have no other choice.

Question 2: What is your restoration philosophy?

I asked my interviewees to discuss their philosophy regarding how we should design and implement our restoration programs and projects. For example, did they see themselves as purists (proceed cautiously, strive for historical accuracy when reestablishing native species and their communities, and so on), pragmatists (proceed rapidly and get over the fallacy of historical accuracy), or somewhere in between? Because some of my colleagues and I had been forced to abandon our initial purity over the years, I thought I was going to find that others had similarly become increasingly pragmatic over time. But I did not find any consistent relationships between people's conservation philosophies and their experience, occupation, employer, education, or expertise. Indeed, the two strongest purists I encountered were also two of the most experienced, broadly trained, and knowledgeable people in the entire state:

Sure, you can get a list of plants and just throw them out there. Often it'll "work" because the plants are accommodating. But that doesn't reflect what's going on in nature in any meaningful sense; that's not restoration, that's gardening. You have to spend a lot of time walking around the place to get a deeper idea of its community structure, then patiently design a plan that takes into account fine-scale processes like moisture, substrate, nutrient and disturbance gradients, locally adapted gene complexes, biogeography, and evolution. The more you know, the deeper you look; the more you see, the better you get.

We need to analyze, fine-tune, and integrate everything so that all of these critically important factors and processes fit together—that's what I'd consider real restoration that's worth doing! We should start by working at small scales in relatively simple systems with a lot of thought and concept building and figure out how to do it right before we attempt to work in more complex systems and at larger spatial scales. But nobody does that out here because everybody is in too much of a hurry; they don't realize how crude what they're calling "restoration" really is. For example, they end up killing things that are eating plants without realizing that some of those things are native and they belong there. We end up building this little garden of threatened and endangered plants that is really a restoration façade. The people that don't know what they're doing shouldn't be allowed out there, yet now they're all gearing up for ever more and larger projects.

The young people and newcomers to Hawai'i are impatient and don't want to listen to those of us who have been out here for many years and see and understand things they don't. For example, they don't stop to think about biological sanitation—they're like surgeons who are so busy preparing for their operations that they forget to wash their hands. They do things like forget to clean their packs or stockpile their fencing materials in the rain forest, and the slugs and weed seeds get all over them, then they fly out to the most remote, pristine ecosystems and create a huge disturbance for the helicopter drops. I've seen that disaster happen over and over again.

Other self-identified purists expressed similar concerns about the words and deeds of what they perceived as the brash, uninformed segment of the conservation community:

It's scary that so many untrained people are out there employing the "let's-just-put-everything-everywhere-and-see-what-survives" approach. They think what they are doing is great, but I don't think they are accomplishing anything. We desperately need more training and oversight, more coordination and standardized protocols for things like genetic purity and seed collecting and follow-up monitoring. But no one seems to have the time for things like that anymore; everyone's out there racing around blindly, just trying to keep their own little pet projects going.

It's all about the value to real conservation. If you do it right—carefully record and document everything you do, use the best science and rigorous genetic protocols, et cetera, then it's high value. I guess it's OK if people want to do some happy/feel-good/backyard community project, but to me that has almost no real value, and in some cases it may ultimately cause more harm than good.

Yet other broadly trained, knowledgeable, veteran members of the Hawaiian conservation community had developed very different philosophies. When I asked one such person what she thought about these kinds of purist perspectives, she nearly hit the roof:

There's really not much usefulness for all that fine-scale nit-picking. We're too much in crisis mode out here; worrying about that stuff is a luxury we don't have. While you're hemming and hawing—boom! We just lost another species. The Hippocratic oath of ecological restoration? I don't buy it one bit—of course we're going to do harm! We've got to quit being such weenies, quit waiting for some new scientific study or new tool to tell us what to do and save us—we need to roll up our sleeves and get to work now!

And all this obsession with genetics—PLEASE! That's the ultimate fallacy, and it demonstrates how far off academia is from the rest of the real world. They're worried about hybridizing related species and contaminating the gene pool? If we've learned anything about Hawaiian evolution, it's that these species are rampant with hybridization, and they're still doing it right now in front of our eyes. "How dare we take a species from one island and hybridize it with one from another?" [they say]. I'd say, "Hallelujah, probably save 'em both!" People are so misguided to think we're going to change the course of evolution in Hawai'i by doing something like that—that kind of thinking is the height of intellectual arrogance.

Given my personal struggles to develop a consistent philosophy to guide my own restoration work, I was secretly gratified when several people initially claimed to be strong purists, then discovered over the course of our interviews that their beliefs were more complex than they had realized. Here's an example:

I'm a purist, pure and simple. If you're going to do restoration, you might as well do it right and go as far as you can before you start sliding down that slippery slope. I guess it might be different if you are doing the restoration to get back some kind of ecosystem services such as watershed protection and erosion control. Or if you can't get the plants to grow that you think should be there. If nonlocal native species will establish, they probably were there once and we just didn't know it, although there are probably exceptions, such as species that evolved in one place just because they came late and didn't have the time to disperse. But I still wouldn't want to use any species in a given restoration project that are restricted to other islands or other regions within the same island because I wouldn't want to destroy that evolutionary patchiness that makes Hawai'i so unique. But I guess that might depend on the specific attributes of the particular species—sometimes we learn more and realize we goofed and need to go back and rethink things and maybe pull some stuff back out of the ground.

Of course, it's good to be flexible because every project is different. For example, some restoration projects have cultural importance, and there's value in that too, although most of the rare stuff doesn't have much cultural importance. When it comes to bird translocations, I'm generally opposed to it unless we're absolutely desperate. And bringing in big alien birds to fill the niches of extinct Hawaiian birds seems ridiculous, but I guess I'd be willing to consider it on a case-by-case basis. I actually wouldn't be opposed to bringing in mainland crows if the last 'alalā goes down—that might add another important piece to the puzzle. I suppose it would make a certain level of sense to bring some species out here and just start the whole evolutionary process over again.

I did find that like me, some people's thinking had shifted toward the pragmatic end of the spectrum as they came to better appreciate the magnitude of Hawai'i's ecological challenges. Some had also come to believe that the concepts of ecological and genetic purity were misguided or at least far more problematic than they had originally realized:

I have definitely become more liberal in terms of taking risks. I'm more willing to try things such as biocontrol and bulldozing than I used to be. I've also come to see that plant communities are not comprised of any fixed, magical combination of species—they're not necessarily coevolved or "balanced," and they can survive and evolve with new players like some of the less obnoxious weeds and plants brought over by the Polynesians.

When I first came out here I was passionately idealistic because I was naive and ignorant and had no idea how bad things really were and what it takes to preserve and restore species and their ecosystems. Now I have less delusions about what this place is going to look like in ten or twenty years, and I just try to get a few simple little things done well and hope that maybe we'll have a chance to revisit the big idealistic vision again a few decades from now. What we really need now most of all is just common sense. We're going to lose the game if we are too pure; that's for total losers because it ain't never gonna happen. And frankly, I guess I'm not even so sure how important being pure is anyway.

Despite all the passion and rhetorical fireworks of the more dogmatic individuals at both ends of the purist/pragmatist continuum, most people saw themselves as being somewhere in the moderate middle. Many also stressed that because the world of ecological restoration is itself so heterogeneous, we should strive to maintain an open-minded, flexible, and tolerant attitude:

We need to be between the extreme purists and the "let's-just-put-everything-everywhere-and-see-what-survives" people. Because everyone has such different conservation visions, we need to have more mutual respect and understanding.

I feel uncomfortable around people with really strong opinions about restoration. The reality is we have to deal with a lot of difficult situations on a case-by-case basis. We shouldn't even try to coordinate or agree on everything, develop standardized protocols, et cetera, because every project is different. We should encourage people to tackle what they find interesting in whatever manner they think is best. That way we'll learn the most and cover the most ground in terms of science, culture, education, and outreach.

Growing up as a liberal in a very conservative place, I was bothered by people who assumed I had the same opinions that they did. So I try to be inclusive and not judge others whose goals and values are very different than mine. When we reach some crossroads where a difficult decision has to be made and we just can't agree, maybe there are times when we should just vote and move on with no hard feelings. Most people can only take so many hours of sitting around talking in circles and arguing anyway, and too much of that tends to frustrate and demoralize the more action-oriented folks.

Question Three: What do you think are the most important things we should do to preserve and restore Hawai'i's natural environment?
In addition to learning more about my interviewees' philosophical orientations, I asked them what they thought we should actually do and how they thought we should do it. Once again, peoples' answers were passionate, contradictory, and unpredictable:

Like Humpty Dumpty, once the egg is cracked, we'll never be able to put it back together again. So we should try to save the best places first; we already know how to do it and we'd get the biggest bang for our buck. I know that the longer you wait, the more difficult restoration becomes, but I still wouldn't support working in heavily degraded areas until all the more intact, high-quality areas are protected first.

Don't give up on the "basket cases" because they're savable. Don't just go after the easy stuff; don't go for the triage model. Nothing is hopeless— shoot for the moon! If we're not putting our efforts into all options and all ecosystems, we're not really trying to save biodiversity.

Coming from the outside, I see the unrelenting fervor in people out here, and I understand it. I cry over the loss of species too, but this extreme emotion and intensity—it takes our work away from rationality, which is not an effective way to do conservation. People want to conserve everything and are resistant to stepping back and imposing a viability filter over our work and consider allocating our resources differently. But my mission is to get people to think about the long-term ecosystem integrity of our projects and let go of the places that really aren't sustainable. I know there are some places and projects that appear to have reversed the tide of degradation, but I think those are at best a temporary blip in time. Like it or not, we need to be ruthless about our priorities and employ the triage model. Should we really be working in places where there are only five individuals of some species left? I know how difficult it is to say no to anything, but we need to consider what else we are going to lose by refusing to give up on those last five individuals.

People come out here and they don't understand how variable things are in Hawai'i—how so often all you have is tiny populations within highly degraded remnant habitats, and how often there's not much else you can do due to politics and biology and limited resources. But these outsiders come in and want us to "think big" and "think systematically" and arrogantly criticize our postage-stamp projects and think they're going to accomplish some grandiose ecosystem-scale project. They just don't understand the critical importance of the partnerships behind those "little projects" and the time and energy it took to get them off the ground.

People are always talking about saving Hawai'i's biodiversity, but the great majority of them only care about birds and plants. If we really wanted to maximize our efforts, we should all be working on arthropods. Their diversity is ten to thirty times greater than the next taxonomic group, and their threats are entirely different. We could preserve almost an entire assemblage of arthropods relatively easily, but no one ever even thinks of doing something like that.

Some argued that the whole "mainstream conservation paradigm" was itself elitist, misguided, and ineffective. Many of these people therefore argued that environmentalists must more effectively engage and support the larger general public:

At first I had a completely biocentric perspective; I was a card-carrying member of the Wilderness Society. But now I realize to be effective, we must include and think about people and culture. I don't even believe in the idea of wilderness anymore—the separation of humans and nature is an artificial and divisive concept. When I think about conservation now, I think more about people and their interaction with the natural world than I do about endangered species and ecosystems.

When I came out of college I was ready to chain myself to trees—got to stop this, save that—but it's not like that anymore. The whole thing is not as pure as I once thought it was; you can't just preserve or restore something because there are always people and cultures related to these things. Take aerial rat baiting, for example. The conservation community was way into that; they discounted local people's concerns and arrogantly tried to "educate" them. To me, that's not a worthy battle. How are you going to look people in the eye and tell them we're going to drop rat poison on them to save some birds?

Most of society doesn't get and can't embrace the big conservation picture, and thus they can't understand the concept that everything is connected. So we should start by working in their neighborhoods and helping people see what's going on back there, but we do the exact opposite by focusing almost exclusively on the remote, pristine areas. Then we tell them that those areas are so precious and fragile that nobody should ever go there except us—no wonder conservation isn't getting anywhere!

Most of what we're doing is masturbation, and I don't use that word flippantly—what we're doing is truly self-stimulation. There are essentially two different camps out here: those gainfully employed by conservation money and everyone else, which of course is the vast majority of Hawai'i's people. Most of the public wouldn't even notice, let alone care, if it all went "poof" tomorrow! So if we're truly interested in preserving what remains of Hawaiian biodiversity, we need to be thinking about social engineering and cultural and societal changes that will actually last in the long run. It's just a matter of time before all of today's conservation infrastructure—the government agencies, the universities, the nonprofit organizations—won't be here, and it'll all be in society's hands. So I'm trying to find ways to bridge the present into that future; I'm trying to find models of conservation that will persist when all of that infrastructure is gone.

I've seen over and over again how critically important education is. Yet while everyone will say they think it's important, no one ever wants to do it! It's always the last thing we tack on at the end of our talks and meetings and papers and grants, the last thing to get funded, and the first to get cut. Conversely, everyone always wants to do more research because everyone argues the things we need to know are so critically important. Well, yeah, they are, but if you don't educate the public, you don't get their support, and then you wind up being unable to do the critical things we already know how to do. So we should be doing things like placing public service announcements on TV, showing a picture of a native bird while they roll the credits after the newscast: "Today's bird is . . ." We should be getting our messages out to the kids on McDonald's tray liners, and selling 'i'iwi *ale to their parents. We have to bring the native species to life and make them stars so that the public can understand what's happening and take some ownership in the efforts to save what's left before it's all gone.*

Question Four: What role does science presently play in guiding ecological restoration in Hawai'i, and what if any changes would you like to see in the future?

This question provoked the most passionate responses from almost all my interviewees. Some people (most but not all of whom were scientists themselves) emphatically stated that science and scientists were of obvious fundamental importance and at least implicitly maintained that rigorous science has been and always will be the driving force behind humanity's overall intellectual and technological progress:

Resource managers often don't realize how much of their knowledge and tools come from science. It's easy for them to take the fruits of science for granted, and say, "Well, we already know all that" or "That's just obvious, commonsense stuff," when the fact is that until scientists did the research, we didn't actually know or understand it at all! I also hear some managers say that we already know enough to act—we know what to do and just need the resources to go out and do it. I think that's bullshit! They don't know nearly enough; in fact, they often have no idea what's really going on out there. Good science can help them prioritize and focus their management activities and reveal some of nature's underlying complexities. If they think they already know everything they need to know, they are ignorant, and that's their problem, not mine!

There are so many things happening out there that are complete mysteries, and we'll never unravel them without real science. We managers can't do it ourselves—we're way too busy and we don't have the necessary training or tools—so we absolutely need to get the scientists in there.

Several people stressed the critical importance of carefully recording, monitoring, and assessing our management activities, as well as various indirect benefits of science and scientists:

We in the conservation community are like pioneers, and we won't know for decades how what we are doing today is going to turn out. But by then, those of us who are doing this stuff mostly won't be around. The people whose work came before us generally left few or no good records about what they did, which has been extremely frustrating. So I always try to make my records scientific and rigorous, because I believe we owe this to the future so they can understand, learn from, and build on what we did.

The science may not have much practical value, but it can be very useful for us to justify and defend why we are doing things like snaring pigs. It will never convince the diehards, but sometimes it helps us convince the funding agencies and landowners and politicians. It's also really helpful when scientists do basic education and outreach work—go into the schools and help little kids and their teachers understand what's happening out here, or give nontechnical talks to the public.

Most people can't understand and aren't even very interested in the technical details of our work as scientists. For instance, my grandmother has no idea what I do, but she's proud of me and thinks what I do must be important. I send her my publications, which she values and shows off, even though of course she and her friends would never actually read any of them! Similarly, because scientists have prestige in our society, we have some power to influence what happens on the ground even though most people really don't understand what we're talking about.

Science has some great stories to tell. A portion of the public just can't get enough of them, especially when they are presented so that nonscientists can understand and appreciate them. When the public grasps the stories about, say, the amazing paths evolution has traveled in Hawai'i or the unique natural history of our native species, they value the natural environment and the efforts to protect and restore it that much more.

Yet many also discussed how scientists and resource managers come from and live within two distinct worlds. As a group, the scientists were less concerned about this issue than the resource managers. Not surprisingly, those who had at least some experience in and understanding of both of these worlds tended to have more empathy for the "other side":

Managers are usually too pessimistic, and scientists too optimistic, about the relevance and importance of science. A lot of the friction also comes from the fact that scientists have more education, greater prestige, better pay, and more freedom than the managers do. They're out there having fun, doing whatever they want to do, while the managers are doing hard, often tedious physical work. No wonder some of them have a chip on their shoulder!

The scientific and resource management communities aren't interfacing well because they represent two different cultures with different needs. The managers need simple, concrete information and tools immediately, but science doesn't work that way—it operates on a slower, one-piece-of-the-puzzle-at-a-time scale. That's not unique to Hawai'i, but because of the urgency of the situation out here, that conflict is intensified, and thus managers often see scientists as fiddling while the house is burning. To some extent it's intractable, although both sides could do a better job communicating with and respecting each other.

Frankly, I don't want to alter my research program to try and solve some narrow applied problem. Science is my passion—it's what I'm trained in, what I'm good at, and what I get paid for—and I think it's important even if it has no immediate practical value.

Scientists always want us to help them get set up, show them where some endangered species is, drop what we are doing to support their work. I admit some science is cool, fascinating stuff, but it's largely irrelevant to us resource managers. And whenever we make a list of the questions we would like to see researched, we're told our issues are too applied or not interesting enough, so nobody takes them on.

My whole existence as a field technician is pretty divorced from science. I'm not against it, but what gets me is when people try to turn our work into a science project. Due to external pressures and funding issues, we're often forced to take the "scientific approach" rather than just directly solving our management problems. Most of the time, our needs are not that nuanced—we've got to build the fences, kill the pigs, get rid of the weeds—who needs a scientific experiment to do that?

A lot of the "conservation scientists" do these tiny little experiments in which they control and rigorously measure everything. Then they publish these studies in their technical academic journals and think it's someone else's job to interpret their results and convert them into management practices. Many of the scientists I know delude themselves into believing the practical implications of their publications are critically important to the real world, but the truth is that neither the managers nor the policy makers have the time, training, or incentives to do that—none of these people ever even read that stuff!

We try to do some experiments and monitoring and analysis of our different techniques and projects. However, if we tried to do "real science," we'd really have to limit the scale of our work. But as a resource manager, I will always sacrifice rigor, replication, data collection, et cetera, for more on-the-ground accomplishments, because that's our mission. Consequently, much of the data we do collect is really messy. There's no continuity or consistency to it, because it's usually collected by different people at different times using different protocols. It's really hard for us to get much out of it, and it's even worse when outside scientists who have never even been out there start analyzing our data and telling us what to do! So mostly we learn by trial and error and following our own intuition and talking to knowledgeable local people.

My background is math and science; I understand why people want to see the quantitative analyses of the effectiveness of our management work. But at this point, we really just need more bodies doing more work. Running all those meticulous and expensive but ultimately irrelevant transects to assess our work just doesn't make any sense.

These kinds of frustrations led some to question the practical relevance and appropriateness of conservation science in general. Some scientists acknowledged that their work probably had little if any applied value, but they often felt pressured to claim otherwise in their grants, publications, and oral presentations. Others believed that scientists and their science can and should become more relevant and helpful and suggested various reforms to facilitate these changes:

Restoration is an art, not a science! Science by its very nature is always incomplete and never actually resolves anything. In all my years in

Hawai'i, I've yet to see a single example in which academic science and all of its theories and models directly helped someone design and implement a resource management strategy or restoration program in the real world. Funding, logistics, politics, partnerships, politics, charisma—those are the things that really drive conservation out here, not science. Restoration is also all about aesthetics and values. What should a restored forest look like? What should we do about all the species brought over by the Polynesians? Should we try to preserve Hawai'i's native biodiversity even if it turns out that we can't scientifically show that native species are necessarily better for things like carbon cycling and watershed management? Science cannot address those kinds of questions.

We're always getting flooded with research applications from scientists who want to go do their research in some remote, pristine place. They always throw in some boilerplate stuff about how valuable their work might be to us managers, and we always laugh when we read those sections because they're total horseshit! Do they think we're that stupid? If it were up to me, I wouldn't let any of them in, because the best we can do for those kinds of places is to prevent all new disturbances—no more researchers, no more surveys, no more flagging tape.

I can't defend the kind of ivory-tower science that dominates the research community out here and is so revered. I read the parts of their papers and hear their talks in which they proclaim how valuable their work is or will be to the folks on the ground, and it's grating. I wish they would at least use language that was a lot more cautious and realistic.

I love doing my research—evolution has given us these big, curious brains that like to ask questions and try to solve obscure puzzles. But I hate writing those "implications for managers" sections everyone always makes you write! I wish we could just skip all that and get back to our science and stop pretending that we're doing something that we all know we're not.

I actually enjoy putting my resource management work into various scientific, academic frameworks. It's a very entertaining pastime, but of course I do it tongue in cheek, because it always ends up being quite a stretch. The real world is so much more complicated and messy than that!

I've been involved with too many research projects in which we gridded up tens of thousands of acres of rain forest. The scientists who pressured us into doing these projects didn't seem to care or even notice that we ended up trampling some really sensitive areas to do that! So we bust up the forest, create all these new avenues for the weeds to come in, lose all that time we could have spent actually doing conservation, and what do we ever get back for our efforts in return? At least in Hawai'i, there should be the equivalent of the "first, do no harm" oath for scientists. It would also help if they found ways to give something back, and I don't mean their research publications! Getting us some funding, contributing some labor towards our projects, giving nontechnical talks, and writing simple pamphlets that help the public understand and support our work—those kinds of things would really help improve our overall relationships.

I feel an ethical obligation to connect my science to conservation and try and make it relevant, but I don't believe that all scientists should necessarily have to do this. Some just don't have that pull, and that's fine, because the pursuit of pure knowledge is a good thing, and they may wind up discovering things that researchers like me never would. But a large portion of at least the government's research programs should be explicitly designed to meet the needs of the management and conservation communities, because that's why they were ostensibly brought out here in the first place. But because these scientists are pressured to get external grants and produce esoteric academic publications, what they actually do is rarely solution oriented—their conclusions are almost always, "We need to do more research!"

There should be a different paradigm for how research science operates out here. For example, medical science might be a much better model. Some of that research is basic science, but even that portion of it is driven by the actual needs of the medical community. Too much of our conventional "conservation" science is driven by academic snobs that look down at the "uneducated" management community and are constantly telling us what to do, rather than asking us what we need. I buy the argument that all this "pure" science will somehow eventually trickle down to and enlighten the lowly world of us practitioners about as much as I buy the theory of trickle-down economics.

Finally, several people stressed that science is only one of many different ways to see, interact with, and learn about the world, and that to more effectively engage the public and garner their support scientists in particular and the larger conservation movement in general need to show more respect for other knowledge and value systems:

> *After I stopped doing academic scientific research and started working with farmers, the way I saw and interacted with the natural world changed drastically. I began to see that science is only one way of perceiving and understanding the world. Sure, it's a really important and interesting perspective, but there are other equally important and interesting points of view as well. I think it's important for all scientists to devote at least a little time and effort towards understanding and appreciating some of these other perspectives, because the vast majority of the people in Hawai'i, and the planet as a whole, do not and probably never will see the world through the lens of science. But if scientists don't understand this and can't relate to any of these other knowledge and value systems, then they and their work will remain alienated from the masses, and their brand of conservation will remain a fringe movement of elites.*

10   NATURE IS DEAD. LONG LIVE NATURE!

In 1989, the environmental journalist Bill McKibben published *The End of Nature*. In this best-selling and influential book, he explained, "By the end of nature I do not mean the end of the world. The rain will still fall and the sun shine, though differently than before. When I say 'nature,' I mean a certain set of human ideas about the world and our place in it." For McKibben, the nail in nature's coffin was human-induced climate change: "We have changed the atmosphere and thus we are changing the weather. By changing the weather, we make every spot on the earth man made and artificial." Consequently, he argued that "An idea, a relationship, can go extinct, just like an animal or a plant. The idea in this case is 'nature,' the separate and wild province, the world apart from man to which he adapted, under whose rules he was born and died."

While there continues to be much debate about McKibben's thesis, many scholars now believe that the concept of nature as some kind of pure, Edenic entity, wholly separate from and "uncontaminated" by humanity, is and in fact should be dead. This idea has been killed by both our increasingly powerful and global effects on the environment as well as our changing perspectives of nature and our past and present relationship to it. In a nutshell, a broad diversity of thinkers has come to the ironic conclusion that the concept of pure nature is itself an unnatural human projection. Their argument is not that the natural world doesn't exist, doesn't have value, or isn't worth fighting for, but rather that we can never conceptualize or experience it in a wholly natural way. As one anthropologist surmised, "Reality is the world as it is perceived by the mind through the medium of the senses. Thus reality in nature is not just what we see, but what we have learned to see."

For example, many of us may see US national parks as pure and noble refuges from civilization and all of its problems. We like to think of the objects and species and phenomena we encounter within these parks as being independent of ourselves and therefore wild and natural. Yet on further reflection, we cannot escape the fact that the designation and "preservation" of such areas were entirely human endeavors that were motivated by a subset of a particular culture with a particular (and historically quite unusual) conceptualization of nature. Moreover, even in the increasingly rare instances when we haven't directly affected the living and nonliving components within these "wildernesses," our perception of them is necessarily unnatural and intimately intertwined with our human world.

Consider the spectacular volcanic eruptions that regularly occur within Hawai'i's Volcanoes National Park. The Native Hawaiians believed that the goddess Pele lived in what is now this park's central crater and that its eruptions were an expression of Pele's longing for her lost true lover. What they saw and felt when they went to this sacred area to offer gifts to their goddess was obviously quite different than the experiences of a modern American volcanologist, which in turn is quite distinct from that of, say, a Japanese ecotourist. The mere fact that I referred to these phenomena as "spectacular volcanic eruptions" reveals at least as much about me and my culture as it does about the eruptions themselves.

Our increasing awareness of the extent to which we have been altering the environment throughout history has led some scholars to conclude that McKibben's concept of uncontaminated wild nature died long before the advent of contemporary human-induced climate change. Virtually everywhere modern researchers have looked, they've found that our ancestors deliberately and unintentionally transformed nature far more than formerly assumed. Ironically, due to their extreme geographic isolation and exceptionally late discovery, the Hawaiian Islands may have been one of the very last strongholds of wild nature. Yet we are now beginning to realize that the sweeping environmental changes that occurred in Hawai'i after humans arrived were not nearly as unique or exceptional as we once thought.

For example, when Christopher Columbus landed on the Island of Hispaniola in 1492, he brought many new species to the New World with him. Much like Captain Cook and his successors did a few hundred years later in Hawai'i (and the Polynesian sailors a thousand years before Cook), Columbus and his successors deliberately introduced animals such as cattle, sheep, and horses, as well as agricultural crops such as sugarcane (originally from New Guinea), wheat from the Middle East, and bananas and coffee from Africa. They also inadvertently introduced many stowaway species such as mosquitoes, cockroaches, honeybees, rodents, dandelions, and African grasses.

What followed was remarkably similar to what happened in Hawai'i: the Old World animals destroyed the undefended native vegetation and facilitated the establishment of noxious alien weeds; Indian mongooses decimated the native snake fauna; native palm, mahogany, and ceiba forests were replaced by Australian acacia, Ethiopian shrubs, and Central American logwood. Today only small fragments of Hispaniola's pre-Columbian forests remain.

As was the case in Hawai'i, the combined effects of these Europeans and their alien species were catastrophic to the native human populations they encountered. By 1548, fewer than five hundred of Hispaniola's once "innumerable" indigenous Taino Indians survived. While human cruelty exacerbated the elimination of this and other native cultures, the Old World epidemic diseases were far more devastating. Throughout the sixteenth and seventeenth centuries, the viruses and bacteria that cause diseases such as smallpox, influenza, tuberculosis, and scarlet fever traveled across the oceans with the

Europeans, swept across the Americas, and ultimately killed at least three-quarters of the people in the Western Hemisphere.

Columbus' voyage may also have set in motion a chain of events that resulted in large-scale, human-induced climate change. Some scientists now believe that as the tens of millions of indigenous Americans died in the wake of the European conquest of the New World, vast areas of cleared land were left untended. Trees then reclaimed much of that cleared land and pulled billions of tons of carbon dioxide out of the atmosphere through the process of photosynthesis. This massive "natural" reforestation event can account for the sudden drop in atmospheric carbon dioxide recorded in Antarctic ice during the sixteenth and seventeenth centuries, which in turn may have diminished the Earth's heat-trapping capacity and caused the centuries of abnormally cool temperatures in Europe following the Middle Ages (Europe's so-called Little Ice Age). Thus humans may have inadvertently but significantly altered a substantial portion of the Earth's climate several centuries before the contemporary global warming that McKibben viewed as the end of nature.

(The above discussion is not intended as a criticism of McKibben's thinking, scholarship, or current leadership within the global movement to help solve the present climate crisis. On the contrary, McKibben is actually one of my heroes, and I am a climate activist myself. Nor do I mean to equate the relatively minor climatic effects of the post-Columbus reforestation events with the massive changes we are causing today. For example, in 2010 alone, humans pumped about nine billion tons of carbon into the atmosphere, an amount roughly equivalent to the total amount of carbon sequestered by the recovering forests in the Americas between 1525 and 1625.)

Ever since the European discovery of the New World, the biological and human compositions of the previously isolated Eastern and Western Hemispheres in particular and the planet as a whole have become increasingly similar. Some have even suggested that Columbus' first voyage to the New World, which started this global homogenization process, marked the beginning of a new biological era they call the Homogenocene.

The past and present interactions of people and the environment in this homogenized brave new world are so complex that determining what is and is not natural has become increasingly difficult and arbitrary. For instance, when the Spanish colonists on Hispaniola imported African plantains in 1516, their native scale insects apparently came along with these plants as well. As was and is the case with some alien species in Hawai'i, because these insects must have left all their competitors, predators, and diseases behind in Africa, their population probably exploded soon after they reached their new home. They also likely benefited from the fact that Hispaniola's native tropical fire ants eat the sweet excrement produced by scale insects and voraciously attack anything that blocks their access to them. When the Spaniards' extensive orange, pomegranate, and cassia plantations on Hispaniola were completely destroyed in 1518 and 1519, they blamed the "infinite number of ants" that swarmed through their houses in such numbers that they ultimately abandoned them and fled in terror. However, scholars now believe the real culprit behind this agricultural and social disaster was the exotic sap-sucking scale insects.

Was the complex—yet by necessity, locally based—agricultural system of Hispaniola's indigenous Taino Indians more natural than the exotic Spanish plantations that replaced it? If the subsequent destruction of these plantations had been caused by a native disease or herbivore, would this have been a more natural process than the destruction caused by the African scale insects?

Modern "back-to-the-land" farmers and gardeners are often motivated in part by a desire to be closer to nature and live a more natural life. But like conservation, agriculture is yet another human invention (though other species such as fungus-raising leaf-cutting ants do it, too) that blurs the boundaries between what is and is not natural. Is it more natural to raise plants and animals whose geographic origins are closer to our homes than those that come from farther away? If so, Americans who wanted to farm as naturally as possible would have to drastically change the kinds of plants and animals they raise.

Similarly, since all of today's agricultural species are derived from many years of intensive, cosmopolitan breeding programs, are the seeds of a native, old-fashioned heirloom variety necessarily more natural than seeds from an exotic hybrid? For example, our culturally infused, locally produced varieties of heirloom "American" tomatoes are probably the distant descendants of one or more species native to

Peru and Ecuador (only one of these species produces edible fruits, and these are about as big as a thumbtack). Ever since Columbus, we have carried tomatoes around the world and radically modified them to suit our various needs and desires. There are now over seven thousand varieties of cultivated tomatoes, but none of these even remotely resembles their wild ancestor.

The purest contemporary organic American farmers must still first either remove most of "nature's" plants and animals from their land or maintain the clearings created by their predecessors. If they are lucky, abundant populations of alien earthworms will help them improve their soil, and nonnative honeybees will pollinate their exotic crops. Even if they decide to use draft horses (originally introduced to the New World by Columbus, although there had been horses in North America until newly arriving humans may have killed them off twelve thousand years ago) instead of tractors, they will still most likely need to purchase several "natural" soil amendments that were manufactured in foreign countries using highly industrialized, fossil fuel–driven processes.

Modern environmentalists must grapple with many similar kinds of physical and philosophical paradoxes. Fundamentally different ideas about what is and is not natural—and to what extent this naturalness matters—underlie much of the tension within Hawai'i's conservation community in particular and the broader environmental community in general. Those who think that nature and humans are essentially polar opposites and that we should preserve and restore the world to as natural a state as possible will obviously strive to erase humanity's past and present footprints from the landscape. This is, to put it mildly, a tough task in Hawai'i, because in addition to all the fiendishly difficult intellectual questions about where humanity ends and nature begins, the physical challenges involved with identifying and removing our footprints remain far beyond the grasp of our present knowledge and technology. People with this conservation philosophy thus tend to be purists who believe we should focus on the preservation of relatively pristine areas and implement restoration projects with extreme caution or not at all. They also tend to be suspicious of people with more liberal philosophies and often conclude that they are doing more harm than good.

On the other hand, an increasing number of people now appear to believe that there is no rigid boundary between people and nature, that naturalness (however defined) should not necessarily be our overarching conservation focus, or both. Consequently, there is a far greater diversity of philosophies among this group than within the purists' camp. Some believe in accomplishing more traditional conservation goals such as preserving native biodiversity, protecting our watersheds, or minimizing soil erosion. Some are more interested in culturally oriented projects such as creating ethnobotanically oriented "canoe gardens" that showcase the species brought to Hawai'i by the Polynesians in their voyaging canoes. Others have more artistic or social motivations—some simply want to enhance the beauty or "feel" of areas they believe are ugly or monotonous. One prominent scholar has argued that in addition to our conventional conservation work, when appropriate, we can and should practice what he calls "inventionist ecology," in which we deliberately create new ecosystems and even species, just as artists create things that presently don't exist.

While not necessarily supporting inventionist ecology, some conservationists believe that rather than attempt to remove, minimize, or hide humanity's presence in the environment, we should celebrate it instead. These people have concluded that the "take only pictures, leave only footprints" model of interacting with nature has failed because it is both practically impossible and because it generates too much despair. They thus see the paradigm of ecological restoration as enabling humanity to "reenter" nature in a more environmentally enlightened and constructive manner. Rather than feeling guilty about having to, say, "artificially" burn the land to restore native Midwestern prairies or manually plant koa to recreate Native Hawaiian rain forests, they argue that we should turn such events into celebratory public festivals. Some maintain that due to all the time and effort we pour into such projects, the resulting restored ecosystems may actually be *more* valuable than their remnant "pristine" counterparts.

Not surprisingly, these more radical philosophies have generated passionate debate within and outside of the scientific and conservation communities. In addition to wrestling with the human/nature relationship, much of this debate has revolved around what are and are not "authentic" restoration projects and human interactions with nature. For example, visitors to America's largest indoor rain forest, the one-and-a-half-acre, eighty-foot tall "Lied Jungle" in Nebraska, can now

> see, touch, smell, hear and become part of the natural rainforest environment while observing animals that are free ranging or contained behind water and rock barriers. Visitors young and old will journey through three different rainforest habitats from Asia, Africa and South America, seeing the lush vegetation and animals native to the area. Along the journey they'll find medicinal plants, giant trees, seven large waterfalls, cliffs and caves. Most of the rockwork, as well as the larger trees, are man made to enhance durability and aesthetics. Materials used to create the man-made rockwork and trees include fiberglass, cement and metal frameworks. All man-made amenities are blended with live plants and animals to almost make the difference unidentifiable. The Lied Jungle's roof, constructed of fiberglass-reinforced plastic, provides natural sunlight to penetrate the rainforest and generate growth. The sense of actually experiencing life in a rainforest is enhanced in a variety of ways. Mechanical devices not native to a jungle, such as air ducts, filters and light fixtures, are hidden in the walls and rocks. Experience jungle life through the eyes of a bird through the jungle canopy or through the eyes of an otter or pygmy hippo by wandering the jungle trail. From either vantage point, the plant life is diverse ranging from towering bamboo and large fig trees to delicate under story plants and orchids. Approximately 90 animal species can be found at all levels, from forest floor to highest canopy.

On the West Coast, nature lovers can go to California's Disneyland and enjoy experiences such as the Big Thunder Mountain Railroad, a "high-speed roller-coaster-style ride on a runaway mine train through a rocky Western wilderness. . . . Based on Bryce Canyon in Utah, the scenery captures the thrill, excitement and beauty of the rugged and untamed landscape of the American West." They can also take their kids to Disney's Redwood Creek Challenge Trail, where "Wilderness Explorer Russell and Dug, the friendly dog from Disney Pixar's *Up,* want to help you earn Wilderness Explorer badges while navigating forested paths, rock climbing, swinging on rope bridges and zipping down suspended slides."

While it is easy for environmentalists to ridicule such "inauthentic fakery," it is even easier for us to forget just how few people will ever set foot in a "real" rain forest, explore a "natural" wilderness area, or see a "wild" large animal. Yet many will visit zoos, botanical gardens, museums, aquaria, city parks, and other popular, Disneyesque attractions. Are the often highly skilled and dedicated staff members who strive to create "authentic" nature experiences in such places doing fundamentally different work than "real" conservationists? Is there such a thing as an authentic ecological restoration project? Are the experiences of people at Disneyland necessarily less meaningful or valuable than their experiences at Hakalau, Auwahi, Limahuli, or Volcanoes National Park?

I do not know the answer to such questions, but I do know how difficult it often was for my colleagues and me to get the larger public excited about the "real" Hawai'i we were striving to preserve and restore. For example, shortly after I started working at the National Tropical Botanical Garden, I discovered that everyone (including me!) was mesmerized by their hundred-acre Allerton Garden on Kaua'i. The creation of this garden began in the mid-twentieth century, when a Chicago philanthropist began the painstaking process of converting what was then a sugarcane plantation into the present "masterpiece of garden paradise and tropical romanticism." From what I observed, many of the people who visit this garden today develop deep emotional and spiritual connections to what they perceive as the real Hawai'i. However, in contrast to the adjacent McBryde Garden, which contains the world's largest living collection of Native Hawaiian plants and several endangered and extinct-in-the-wild specimens, nothing in the Allerton Garden is native to Hawai'i. Yet I repeatedly found that—compared to the splendor and charms of the Allerton Garden—getting the public excited about this relatively drab and scruffy native garden was a tough and often impossible task.

Similarly, most of my mainland visitors were far more interested in and moved by Hawai'i's picturesque yet "artificial" landscapes—such as the expansive cowboy country of the Big Island's Parker Ranch (created by centuries of intensive cattle grazing) and the stark, Grand Canyon–like quality of Kaua'i's Waimea Canyon (significantly modified by intensive feral goat browsing)—than the remnant "natural" areas I dragged them to. They also often found some of the islands' most noxious alien species to be far more captivating than their non-showy natives.

For example, *Miconia calvescens*, the "Green Cancer," is Hawai'i's poster child of invasive plants. This fast-growing tree from Latin America produces huge, oval-shaped leaves with striking purple undersides. It has already wreaked ecological and social havoc in about 70 percent of Tahiti's native rain forests, and because many fear it could similarly devastate vast areas of Hawai'i, aggressive land- and air-based containment efforts have been launched across the state. Yet when I took my visitors on the Onomea Bay Scenic Drive along the Big Island's Hamakua Coast, most were far more impressed by the exotic beauty of the "purple plant" than they were by my efforts to explain its ecological destructiveness or point out the uniqueness and tenacity of the few remaining native plants that miraculously still inhabit this region of the island.

As my interviews of a broad cross-section of Hawai'i's environmental community revealed, there is an enormous diversity of perspectives about how to conceptualize and manage the islands' natural resources even within this ostensibly like-minded group. Some stridently believe we should fight every last noxious weed and bug and never give up on any native species or ecosystem. Others argue with equal fervor that such thinking is naive, and if we don't systematically prioritize our efforts we are going to lose everything. Indeed, even the US Forest Service recently announced that restoring Hawai'i's tropical forests is "no longer financially or physically feasible."

The magnitude of its present environmental challenges and the intensity of the debate over what to do about them are two more pieces in Hawai'i's conservation puzzle that are no longer as unique as they once may have been. For example, after twenty years of fighting nonnative species in the Galápagos, to the extreme chagrin of some environmentalists, the leader of these islands' restoration team recently stated, "As scientists and conservationists, we need to recognize that we've failed: Galápagos will never be pristine.... It's time to embrace the aliens." He also stressed that with thirty thousand people now living in the Galápagos, effective conservation must increasingly dovetail with local economic activities such as nonnative forest and coffee plantations.

Similarly, a group of prominent scientists recently concluded that "Escalating global change is resulting in widespread no-analogue environments and novel ecosystems that render traditional goals unachievable.... We should not... give the impression that ways can be found to either hold or turn back the clock and preserve or recreate imagined Edens." A growing band of ecologists and environmentalists now maintain that these so-called novel ecosystems—communities comprised of mixtures of native and alien species that never existed in the past—may have their own important yet generally underappreciated value. Some further argue that because nature is often surprisingly resilient, at least some native species will eventually adapt to and even thrive within these novel ecosystems in the same way that they responded to the drastic "natural" changes in their evolutionary past.

A good example of this scenario in Hawai'i is the *pueo,* or Hawaiian owl. We know from the fossil record that most of the main islands once had their own distinct species of long-legged, short-winged stilt-owls. Soon after reaching this archipelago, these predatory birds must have shifted from eating vertebrate prey (mostly small mammals and reptiles) to a diet of mostly or only birds (probably Hawaiian honeycreepers, flightless rails, and perhaps some other forest birds and insects). But none of these owls was able to adapt to the islands' rapidly changing posthuman environment, and today all of them are extinct. In contrast, no *pueo* fossils have ever been discovered in Hawai'i, presumably because they could not establish in the archipelago's prehuman environment of few grasslands and no rodents. But today these owls inhabit all the major islands, and because they are active during the day they are often spotted foraging for rodents and large insects in highly degraded, novel-ecosystem-type landscapes such as the Big Island's Parker Ranch. (The fate of the islands' only other surviving native bird of prey, the *'io* or Hawaiian hawk, remains unclear; at present it is federally endangered and extant only on the Big Island.)

Yet another increasingly popular view of today's complex mosaics of "natural" and novel ecosystems is that they are in essence a series of gardens. Indeed, some argue that the discipline of restoration ecology is itself just another form of agriculture, and that its practitioners do more or less the same thing as conventional farmers and gardeners. Under this model, Disney's "natural" attractions, botanical gardens and zoos, and national parks are all just different points on a continuum of increasingly wild but nevertheless human-created gardens. Some proponents of this paradigm thus believe that rather than endlessly struggle with the intractable boundaries of the human/nature dichotomy or pretend that human-free areas still exist, we should consciously manage our gardens to accomplish explicitly stated objectives such as entertainment, education, research, resource extraction, and conservation. Some even advocate a rewilding process that involves increasing the size and diversity of our "wildland gardens" by connecting them and establishing populations of ecologically important large mammals such as wolves, bison, and grizzly bears. One "Pleistocene rewilding" proposal envisions the potential benefits of reintroducing close relatives of formerly extant species such as camels and elephants back to carefully selected regions of North America.

How such a rewilding process might unfold in Hawai'i depends, of course, on who you ask. Some advocate limiting ourselves to reestablishing native animals and plants to only those areas in which we are confident they previously existed, though there is disagreement within this camp on the validity of posthuman distributions (does it count if a species once existed on a given island because the Polynesians or early Europeans put it there?). Some would be willing to go one step further by establishing species in areas outside of their historic distributions as long as these introductions were restricted to the species' native islands. Others believe we could and should take increasingly bolder steps, such as introducing native species to Hawaiian Islands that (as far as we know) they never colonized on their own; introducing Native Hawaiian species to novel areas outside of this archipelago; and establishing nonnative species in Hawai'i to, say, carry out ecological services such as pollination or seed dispersal that were formerly performed by now extinct native animals.

As horrifying or exciting as these proposals might be, several of my interviewees thought that rather than waste our time arguing about them, we should focus our efforts on implementing some of the effective yet underutilized management strategies we already have. For example, virtually everyone within Hawai'i's conservation community acknowledges the overarching importance of keeping ungulates out of the islands' remaining natural areas. As the projects featured in this book demonstrate, we already know how to do this, and we know that even badly degraded areas can be at least partially restored once these animals are removed. Some thus believe that if politics and limited resources are the main obstacles preventing us from doing more surefire ungulate removal/restoration–type projects, why open the can of worms that would inevitably result from lobbying for far more divisive projects such as introducing exotic birds to replace our extinct avifauna?

But, wouldn't it be great if we could wave our magic wands and get rid of all those ungulates without the expensive, crude, and controversial helicopters, guns, snares, and poisons and surgically eradicate all those noxious alien diseases and bugs and weeds while we're at it? Indeed, several interviewees told me that one of their major goals is simply to slow the rate of ecological degradation and hold the line until new scientific discoveries enable us to accomplish things we can't even imagine today.

The closest scientists have ever come to waving a magic wand at large-scale ecological problems was probably releasing a few miraculously successful biological control agents. Yet as the battle to eradicate the gorse surrounding the Hakalau Forest National Wildlife Refuge illustrated, what if any role biological control should play in shaping Hawai'i's ecological future remains a highly contentious subject. Opponents often start by highlighting the infamous examples of biocontrol backfire in the islands such as the introduction of mongooses (to control rodents) and rosy cannibal snails (to control the giant African snails); in both cases, these exotic species failed to control their targets and became noxious pests themselves. They also argue that because most biocontrol agents do little except perhaps hurt nontarget native species, their ecological risks and economic costs far outweigh their potential benefits (it often requires well over a million dollars and a decade of research to release a single biocontrol agent).

Proponents insist that the science and practice of biocontrol has progressed enormously since the bad old days, and thus disasters such as the mongoose could never happen in today's more rigorous and regulated world. They also believe that biocontrol is our only real hope of ever controlling some of Hawai'i's most entrenched and otherwise intractable alien pests. Their biocontrol poster child is the famous prickly pear cactus–eating moth in the genus *Cactoblastis* (my all-time favorite scientific name—whoever came up with it must have watched lots of Bugs Bunny/Roadrunner cartoons).

What the *Cactoblastis* moth did to the invasive prickly pear cactus in Australia is the world's most impressive example of the potential power of biocontrol. In 1925, about three thousand eggs derived from *Cactoblastis* larvae collected from its native range in Argentina were shipped to Australia. Over the next nine years, over 2.7 billion eggs were mass-reared from this initial shipment and placed on prickly pear infestations in Queensland and New South Wales. At the start of this program, about 60 million acres of Australia were infested by prickly pear cactus; half of this area was so bad that people had essentially abandoned it. Yet two representative quotes from local observers of the *Cactoblastis* release program illustrate just how quickly and effectively this moth did its job:

> The prickly pear territory has been transformed as though by magic from a wilderness to a scene of prosperous endeavour.

> The most optimistic scientific opinion could not have foreseen the extent and completeness of the destruction. The spectacle of mile after mile of heavy prickly pear growth collapsing *en masse* and disappearing in the short space of a few years did not appear to fall within the bounds of possibility.

Today the *Cactoblastis* Memorial Hall and *Cactoblastis* Cairn in Queensland attest to the public's lasting gratitude to the moth and its handlers for delivering them from their "Green Hell."

However, for reasons that remain unclear, the success of *Cactoblastis* release programs outside of Australia has generally been mixed and far less spectacular. And to the extreme consternation of conservationists, some native prickly pear cacti in the southern United States, Mexico, and Central America are now threatened by resident *Cactoblastis* populations; whether these moths got there by deliberate (yet unauthorized) releases, accidental introductions, or natural dispersal from other areas in which they were intentionally released also remains unclear and contentious.

To date there have been no unequivocal biological control success stories in Hawai'i, though some claim that a few existing pests, such as prickly pear cactus itself, would be a lot worse if biocontrol agents hadn't been released against them in the past. But because there have been few careful studies of the *Cactoblastis* releases in Hawai'i, it is difficult to know how much credit this moth really deserves for the dramatic decline of some of the islands' formerly vast prickly pear infestations that occurred around the middle of the twentieth century.

Yet even if we had the necessary consensus and resources to pursue a vigorous biological control program, we'd still be stuck with the fact that some of the most noxious ecological pests are economically or aesthetically valuable to other individuals and special interest groups. For instance, some of the world's most destructive weeds are invasive grasses. In fact, a large portion of my own work in Hawai'i focused on how best to control the fire-promoting blanket of African fountain grass that dominates vast sections of the arid west side of the Big Island. Yet although this unpalatable species is also despised by the island's ranching community (but not the horticulturalists who plant it around homes and businesses to help them blend in with the surrounding "natural" landscape), they would never support using biocontrol agents against it because another closely related African species (kikuyu grass, the same "weed" fought by Jack Jeffrey at Hakalau and Art Medeiros at Auwahi) is an important component of their cattle's diet. Moreover, because so many of our most important food crops are grasses, the broader agricultural community is understandably wary of any biocontrol program that targets this taxonomic group.

Nevertheless, despite such political and economic complications, several of my colleagues and I dreamed of discovering some kind of intellectual or practical silver bullet that could dramatically improve our conservation efforts. Because we knew that progress had often been painfully slow during the early primitive stages of many other scientific disciplines such as physics, we hoped that our young disciplines of restoration ecology and conservation biology might similarly one day blossom into mature sciences with as much unifying theory and practical power as modern physics.

But the more time I spent in the trenches of Hawaiian conservation, the more I began to realize that conservation is not rocket science—it's much harder! In contrast to the relative simplicity of understanding and manipulating the inanimate physical forces involved with building rockets, conservationists must deal with the infinitely greater complexity of interacting living species and the seemingly intractable world of human desires. Thus even if we ever developed a potent arsenal of conceptual or technical silver bullets, we would still have to grapple with our radically different philosophies about such fundamentally important underlying issues as the human/nature relationship. Consequently, we would still have to argue among ourselves and with all the other stakeholders about whether a particular piece of land should be turned into, say, a prehuman wilderness, a prehistoric botanical park replete with interpretive Polynesian canoe gardens, or a modern shopping mall.

Yet people seem to increasingly view science and its technologies as at least metaphorical magic wands and silver bullets. They fervently believe in science as if it was their religion. They want its high priests to use their collective brilliance to distinguish right from wrong and tell us what to do.

I love science; I have devoted most of my life to studying, performing, and teaching it. Science is an undeniably powerful and often beautiful methodology, and if I were king there'd be a lot more of it. But it is not magic, and science and scientists should not be worshipped. Moreover, because science is a tool rather than an ideology or religion, it cannot tell us what to do or believe in, and it cannot resolve our philosophical and practical differences. Even in the rare instances when people agree to base their decisions on a particular research program, they can and often do argue over the best way to perform, interpret, and apply this research in the messy real world. Thus expecting scientists by themselves to solve complex environmental problems such as how to manage a natural area is as misguided as expecting architects to solve complex social problems such as homelessness or political scientists to tell us for whom to vote.

When a relevant human community agrees to tackle a particular technical challenge, science may indeed be the best or even only viable tool to help us investigate and hopefully solve the problem. For example, suppose we wanted to send a rocket to Saturn or develop a more effective birth control pill. Most of us would agree that we should employ a team of engineers, biochemists, and medical doctors to scientifically study these problems and provide technical advice and solutions.

But science is not the right tool when the relevant questions and variables are not technical or quantifiable, and even when they are, science by itself cannot tell us what to do. Thus in this case, whether or not we *should* go to Saturn or develop a new birth control pill are questions for our larger society. Similarly, while scientists can and should answer specific technical questions related to these problems (What is the probability that this rocket will reach Saturn safely? How might this birth control pill affect the long-term health of the women who use it?), they cannot resolve any of the nontechnical but fundamentally important questions that swirl around such projects (How much money should we spend on space exploration? Should we provide free birth control technology to anyone who wants it?).

Yet many people, special interest groups, and politicians are increasingly demanding science-based solutions to an ever-widening array of complex social issues: Science-based education! Science-based health care! Science-based natural resource management! But what in this context does "science-based" actually mean? Proceeding in a logical, deliberate, and methodical manner? Utilizing "objective" evidence? Collecting lots of quantitative data to test and refine a series of hypotheses? Relying on the expertise and judgments of scientists and other highly trained and educated professionals? The reality is that highly trained and educated scientists and other professionals who employ all of these techniques often reach radically different conclusions about, say, what we should teach our kids, whether or not we should reform our health care system, and how we should manage our natural resources.

In addition to different ideas about the relationship between humans and nature, fundamentally different ideas about the role that science can and should play in natural resource management issues comprise another important divisive factor within and among Hawai'i's environmentally related communities. Indeed, my interviews revealed an incredible diversity of different conservation philosophies and battle plans. Some rejected the entire paradigm of science-driven conservation and deeply resented what they saw as the inappropriate and counterproductive intrusions of scientists and their "scientific authoritarianism." Others were convinced that their personal philosophies and on-the-ground strategies were supported and justified by science and were angered by what they saw as the destructive meddling of uneducated or biased individuals, organizations, and special interest groups. Many of these people further argued that natural resource management issues in general should be analyzed and resolved primarily by scientists and other relevant and credentialed experts using "the best available science."

My advice to the people in this latter camp is to be careful what you wish for. First, many nonscientists—and even some scientists themselves—do not realize just how diverse and divided the scientific community often is. We are informed and motivated by a complex mixture of rationality, emotion, altruism, selfishness, ego, ideology, and aesthetics. Even scientists in the same subdiscipline may passionately disagree on narrow technical details within their own areas of expertise, broader professional topics such as what constitutes good science, and of course more complex issues that lie outside the sciences. In other words, scientists and other highly educated professionals are just like everybody else.

Second, even if and when the experts do agree with each other, they are often wrong, or at least not any smarter than the rest of us. Research has repeatedly demonstrated that the predictions of credentialed professionals are no better than those made by the general public. In fact, there appears to be an overall inverse relationship between the accuracy of an expert's prediction and his or her self-confidence, fame, and (beyond a certain minimum level) depth of knowledge.

Several studies have even shown that expertise and experience do not make people better at interpreting scientific data. For example, one study found that when data from a test used to diagnose brain damage were shown to clinical psychologists and their secretaries, the accuracy of the psychologists' diagnoses was no better than the secretaries'.

Third, some hard-nosed scientists have ironically begun to discover that the scientific method itself isn't as infallible as they had assumed. For instance, the test of replicability is widely considered the foundation of modern research; for the scientific community to be convinced that a particular theory or empirical result is real and true, different scientists in different places must independently confirm them and publish their results in a rigorous, refereed journal. Yet in recent years, many ostensibly well replicated and widely accepted findings across a broad spectrum of scientific disciplines have started to unravel. While the reasons behind this trend remain unclear, some believe it is being driven largely by publication bias (science journals tend to publish positive studies that support their hypotheses and reject studies that fail to confirm them) and selective reporting (scientists tend to devote most of their time and effort to data sets that support their prior beliefs and are most likely to get published).

Commenting on what he argues are widespread distortions in the biomedical research literature, one prominent epidemiologist explained, "It feels good to validate a hypothesis. It feels even better when you've got a financial interest in the idea or your career depends on it." In a recent review article, an evolutionary biologist similarly observed, "We cannot escape the troubling conclusion that some—perhaps many—cherished generalities are at best exaggerated in their biological significance and at worst a collective illusion nurtured by strong a-priori beliefs often repeated." Due to a combination of the often intense pressures on scientists to publish something exciting and the speed and power of modern information technology, the life span of many scientific "truths" has become almost comically short. As one science journalist recently concluded, "We like to pretend that our experiments define the truth for us. But that's often not the case. Just because an idea is true doesn't mean it can be proved. And just because an idea is proved doesn't mean it's true. When the experiments are done, we still have to choose what to believe."

Even a cursory glance at history reveals that the track record of "science-based" policies and programs has been mixed at best. Moreover, when scientists and other experts get mixed up with politics, money, and powerful special interests, they can be used to legitimize and justify some absolutely horrible ideas and actions. The seemingly progressive and even altruistic conservation movement has by no means been immune to this problem. An informative case study of this phenomenon is our long and generally unacknowledged relationship with racism and eugenics.

In the late 1800s and early 1900s, the great world's fairs in the United States attracted a substantial portion of the country's entire population and many of its most important civic leaders. Supported by numerous theories and empirical work developed in the nation's finest universities, natural history museums, learned societies, and scientific literature, these fairs displayed living Native Americans as "hideous brutes fit for extinction" to crowds of jeering spectators. At the Louisiana Purchase Exposition in Saint Louis in 1904, which President Teddy Roosevelt toured approvingly with his daughter, scientists measured the physical features and intellectual intelligence of a diverse collection of the world's indigenous peoples and presented these data as more proof of the superiority of whites.

In his famous 1910 "New Nationalism" speech, Roosevelt passionately argued for a new era of sweeping conservation, but he explicitly placed the "morality" and "patriotic duty" of this conservation into a racial framework that linked the richness of America's natural resources with the superiority of its dominant white race. This speech had been written by Roosevelt's friend Gifford Pinchot, who had been his conservation chief for the two terms of his presidency. In 1909, Pinchot had solicited contributions from leading scientists for a three-volume National Conservation Commission report that Roosevelt deemed "one of the most fundamentally important documents ever laid before the American people." Along with a series of idealistic recommendations such as an end to air and water pollution, one of these volumes called for "eugenics, or hygiene for future generations," with forced sterilization of undesirables and eugenically favored (white) marriages. Another volume blamed the demise of American Indians and Native Hawaiians on their own sexual immorality.

Because at that time a lot of money and power lay behind eugenics, supporting this movement was undoubtedly politically expedient. Yet because Roosevelt and Pinchot genuinely believed in science and expertise, their views were also shaped by the many eminent social and natural scientists who supported eugenics, including its creator, the anthropologist Francis Dalton (Darwin's cousin), and the groundbreaking conservationist Madison Grant. Grant, who was an influential friend of Roosevelt's, wrote one the most important eugenics books that Hitler later called his "Bible."

While times have obviously changed, some believe that its eugenics past still unwittingly haunts conservationists and our larger movement. For example, the dominant scientific view in the mid-twentieth century was that Hawai'i's native species and ecosystems were "inferior" and should therefore be "invigorated" by stronger and fitter alien species. Consequently, the territory and later state of Hawai'i used airplanes to aerially seed remote, relatively intact native areas with what today are considered some of the most noxious and intractable alien weeds.

This racist and elitist history may partially explain why many modern indigenous groups, minorities, and poor people are often hostile to environmentalists (whose ranks are dominated by relatively white, affluent people) despite often sharing their goals. A prominent scholar has even suggested that one of the unspoken principles of modern conservationists and their "ecological religion" is that "An (unelected) community of environmentally conscious, morally refined, sober, devout, humble, and self-denying ecological Brahmins should interpret to the masses the will of Nature and direct them accordingly."

These kinds of power struggles and socioeconomic tensions were undoubtedly important underlying factors in the infamous "Chicago restoration controversy" that began in 1996. Called a "loss of innocence" by one environmental philosopher, this conflict revolved around a ten-year, $11.6-million Natural Areas Management Program plan to restore thousands of acres in the larger Chicago area back to the oak savanna/tallgrass prairies that had existed in this region prior to European settlements (and had probably been created by the intentional fire-setting practices of the indigenous people who preceded them).

Prior to this ambitious restoration plan, natural resource management efforts in and around Chicago had largely focused on reforestation and fire suppression or had employed a hands-off philosophy. Not surprisingly, conflict erupted when officials began setting fires, cutting trees, and spraying herbicides. Local citizens, new tree-protecting environmental groups, and animal rights organizations protested the management activities conducted in their local preserves by the "so-called experts in the name of conservation." Volunteer restoration groups, professional resource managers, established environmental groups, scientists, and other members of a recently developed coalition of organizations called Chicago Wilderness fought back. Many of these restorationists viewed their opponents as misinformed, unscientific, and overly emotional NIMBYs (not in my backyard) and called for public environmental education campaigns to bring these people onboard.

Here are a few quotes from the Chicago "antirestorationists" that may sound all too familiar to those who have experienced Hawaiian conservation conflicts. These quotes were collected by a sociologist who later argued that this and other environmental conflicts are mainly about "the structure and process of social and political relationships—who's got the power and who doesn't":

> I have no objections to the purported purposes of the Biodiversity Council [Chicago Wilderness] but I do have serious concerns about the apparent exclusionary nature of its membership policy. . . . [If] the purpose of the organization is to amass a number of organizations, all of which share one limited vision of land and wildlife management and to exclude those organizations which do not share that viewpoint, then the "Biodiversity Council" could be fairly characterized as a *political action committee.*

> . . . 40 years of walking my dogs and taking my children in these woods. I can tell you when the trillium comes up or how many more or less oak trees are here and which birds are declining. . . . Now these people come here and tell me I don't know what is going on in my forest preserve?!

How do they know what my version of nature is? . . . Who's to say what is good . . . who can judge?

A true scientist understands the tenuous hold on truth that science has—that it is no better or worse than any other system of truth finding. In fact, science is much better off if they stay away from the "truth" and try to stick to the facts.

I got angry and I stood up and when I was finally called on I said, "What's so special about prairies?!" The young man next to me said, "Shut up and sit down! You don't know what you're talking about!" Well, that galvanized me! Nobody tells me to shut up and sit down. I am a citizen. I have a right to say what I feel. . . . I do not think the public is as stupid as [the experts] think. . . . They feel that the public is too dumb to understand how great this is.

All of the scientists and environmentalists I know are far too polite to ever tell someone in a public meeting to shut up and sit down. But when we think some fool is babbling on about something they don't understand, obscuring the facts with his blatant ignorance, and poisoning our chances of ever doing the right thing, many of us may secretly wish that someone else would tell him to shut up and go back to school (I know I have). We also tend to love our rarified scientific theories that eloquently reveal the truth, simplify complex phenomena, and show us how to properly perceive, understand, and interact with the natural world. Some of us spend a lot of time trying to come up with such theories ourselves, and we collectively revere those who devise the ones that best capture our hearts and minds.

Yet history has generally not been kind to these scientists and their overarching theories; inevitably, it turns out that things were far more complicated than they and their followers realized. Their grand visions usually turn out to have been influenced by the peculiarities of their class and culture and personal biases in ways that were largely invisible in their time but painfully obvious to us in ours. In hindsight, we see how their theories may have ultimately misled and constrained people at least as much as they informed and guided them.

Ironically, the more complex and severe our environmental problems become, the more we seem to want and even demand simplistic models to explain these problems and show us the best or only way to solve them. Indeed, as "the environment" began to emerge as a hot topic in the late 1960s and 1970s in the United States, several scientists became famous in part by confidently espousing their theories of the true underlying causes of our environmental crisis and how to fix them. For example, Paul Ehrlich and Garret Hardin, two American ecologists, emerged as the leading gurus of the dogma that "population growth is the problem; population control is the solution," and their writings—especially Ehrlich's *The Population Bomb* and Hardin's *The Tragedy of the Commons,* both published in 1968—became the bibles of this school of thought.

Shortly thereafter, the American biologist Barry Commoner appeared on the cover of *Time* magazine as the "Paul Revere of Ecology" and later explained in his own best-selling book that our economy must be restructured to conform to the "unbending laws of ecology." He denounced Ehrlich and Hardin's ideas and accused them of reading into the environmental crisis "whatever conclusions their own beliefs . . . suggested." Ehrlich fired back by arguing that Commoner's work relied on "biased selection of data, unconventional definitions, numerical sleight of hand, and bad ecology" and maintained that his "eco-socialism" was deluding the public by offering them an "uncomplicated, socially comfortable, and hence seductive" solution. This and other related debates raged on over the following decades and into the present as others jumped into the fray, made a name for themselves by leveling similarly harsh criticisms at their competitor's theories, and revealing the real source of and solution to our environmental woes.

My own overarching brilliant theory (guaranteed never to make me famous) is that the world is rarely if ever that black and white. It always seems to turn out that there was at least some wisdom contained within all of the competing theories, as well as important factors that were not addressed by any of them. Thus in this case, most contemporary environmentalists would agree that *both* population growth and the destructive nature of American capitalism are important, but so are a suite of other factors that were ignored or discounted by both sides of that debate. Yet our continuing tendency to promote the particular models that support our own ideologies and tear down the ones that don't foments divisive internal struggles and undermines our collective ability to build the broader coalitions that could actually solve or at least mitigate some of our most pressing environmental problems.

No environmental challenge is more in need of this kind of tolerant pluralism than our ongoing efforts to preserve, restore, and perhaps even create nature (in all its many forms and meanings) within the crucible of modern Hawai'i. Over my years of traveling across these islands, observing and participating in many on-the-ground projects and talking to individuals working on various pieces of the larger conservation puzzle, I was often struck by most people's sincerity and dedication to the cause as they saw it. Despite their radically different perspectives, philosophies, backgrounds, and responsibilities, all of the people I interviewed seemed to truly believe in their work and regularly went far beyond their actual job descriptions and meager paychecks to try and accomplish the things they felt were worth doing.

Sadly, however, I was also struck by the intensity of the frustration and distrust that some individuals and factions had for one another. On several occasions, person X would say, "I don't know what's happened to person Y! She used to really get it and do great work, but ever since she started working for institution Z, she's slowly but steadily gone to the dark side, and now I think she's doing more harm than good." I could usually understand where person X was coming from and empathize with his frustrations, but then I would inevitably talk to person Y, who would proceed to tell me how sad it was that person X had gone to the dark side and was now doing more harm than good.

To be sure, I think there *are* a few people out there who may be doing more harm than good; indeed, some friends and I used to enjoy arguing over whose assassination would yield the greatest immediate conservation gains for Hawai'i. But these people were almost always the upper-level professional bureaucrats who were more interested in the advancement of their own careers than effective conservation (with a few notable exceptions, I didn't bother to interview such people because I knew they would not or could not speak candidly). Yet perhaps because there is so little money and opportunity for professional advancement within the more rank-and-file world of the people I spoke with, I almost never found cause to question their integrity. Consequently, I often wound up concluding that *both* persons X and Y were doing good work, and that their divergent philosophies and approaches each made sense within the context of their specific jobs and intellectual and ecological frameworks.

Conservation is a kaleidoscope of interacting species, ecosystems, people, and cultures that is fueled by a rich mixture of values, aesthetics, science, art, philosophy, ego, and shifting alliances among government agencies, private organizations, special interest groups, local communities, and charismatic individuals. It often makes a mockery out of our best attempts to tame and corral it with unifying theories, formal reductionist science, and rigid philosophical and practical frameworks. Indeed, the theory and practice of conservation has provided me with more opportunities than I care to admit to have my own deeply held convictions crushed by the complexity of the real world. I've also lived to regret crusading for things that at the time I was absolutely certain were right. Now I'm inclined to agree with the general conclusion reached by an anthropologist after years of struggling to reconcile the opposing perspectives of Western science and the indigenous culture he was studying: "Certainty is for those who have learned and believed only one truth."

Yet another humbling feature of conservation is that even when we are as certain as we can be that something is right today, there is no guarantee that it will be right tomorrow. I realized this one day after talking with a Native Hawaiian woman about the historical importance of pigs in Polynesian cultures and how difficult it was for a haole environmentalist like me to understand that so many contemporary Native Hawaiians still passionately believe that feral pig populations actually *help* the islands' native forests. She empathized with how hard it is for all of us to let go of things that were once personally and culturally important. (Contrary to what many nonscientists believe, this problem is at least as prevalent within the scientific community as it is outside of it. As the great German physicist Max Planck once noted, "A new scientific truth does not triumph by convincing its opponents and making them see the light, but rather because its opponents eventually die.") The woman continued thoughtfully,

> In this case, although the Polynesian pigs were kept as a cultured, domesticated animal, when they did get out, some of the old folks used to talk about how the pigs were good for the forest because they *huli* [turn over] in the ground, and then the seeds fall in their wallows, and then the forest flourishes. Well, that might have worked in 1610, when the native ecosystems were so much more wholesome than they are today, but it's not working now because we have all these new invasive species here. So deeply held cultural truisms that once may well have been true do not necessarily always remain true forever. As we weave our way through the shifting sands and moving tides, both individually and as a community, we collectively need to change our wisdom and values.

Despite all our differences and the often extreme intellectual, socioeconomic, and physical challenges inherent to real-world conservation in Hawai'i and elsewhere, we need not fall into the abyss of division and despair and paralysis. We probably never will or even should necessarily agree on how to conceptualize, prioritize, implement, and refine our work. But as the people and programs profiled in this book demonstrate, we can accomplish great things that are worth doing on multiple levels when we pragmatically search for our own solutions with humility, flexibility, and tolerance. I hope that we accomplish many more of these kinds of projects, and I hope they wind up mirroring our planet's natural and cultural diversity. Finally, rather than sitting down and shutting up, I hope that ever more people will, like that person in Chicago, stand up, speak their minds, and get involved.

BIBLIOGRAPHY

Introduction: **Restoring a Rainbow**

Burney, David A. *Back to the Future in the Caves of Kauaʻi: A Scientist's Adventures in the Dark.* New Haven: Yale University Press, 2010.
Burney, David A., and Timothy F. Flannery. "Fifty Millennia of Catastrophic Extinctions after Human Contact." *Trends in Ecology and Evolution* 20 (2005): 395–401.
Clague, David A. "Geology." In *Atlas of Hawaiʻi,* 3rd ed., edited by Sonia P. Juvik and James O. Juvik, 37–46. Honolulu: University of Hawaiʻi Press, 1998.
Cronon, William. *Changes in the Land: Indians, Colonists, and the Ecology of New England.* New York: Hill and Wang, 1983.
———. "Foreword to the Paperback Edition." In *Uncommon Ground: Rethinking the Human Place in Nature,* edited by William Cronon, 19–22. New York: W. W. Norton, 1996.
Cuddihy, Linda W., and Charles P. Stone. *Alteration of Native Hawaiian Vegetation.* Honolulu: University of Hawaiʻi Cooperative National Park Resources Studies Unit, 1990.
Kāne, Herb K. *Ancient Hawaiʻi.* Captain Cook, HI: Kawainui Press, 1997.
Kirch, Patrick V., and Terry L. Hunt, eds. *Historical Ecology in the Pacific Islands.* New Haven, CT: Yale University Press, 1997.

Mann, Charles C. *1491*. New York: Knopf, 2005.

———. *1493: Uncovering the New World Columbus Created*. New York: Knopf, 2011.

Olson, Storrs L., and Helen F. James. "The Role of Polynesians in the Extinction of the Avifauna of the Hawaiian Islands." In *Quaternary Extinctions: A Prehistoric Revolution*, edited by Paul S. Martin and Richard G. Klein, 768–780. Tucson: University of Arizona Press, 1984.

Plant Extinction Prevention Program of Hawai'i Web site. Accessed May 12, 2011: http://pepphi.org/.

Polynesian Voyaging Society Web site. Accessed August 4, 2011: http://pvs.kcc.Hawai'i.edu/.

"Po'ouli Fact Sheet." Accessed October 23, 2011: http://www.state.hi.us/dlnr/dofaw/pubs/endgrspp/.

Powell, Alvin. *The Race to Save the World's Rarest Bird: The Discovery and Death of the Po'ouli*. Mechanicsburg, PA: Stackpole Books, 2008.

Schmitt, Robert C. "Population." In *Atlas of Hawai'i*, 3rd ed., edited by Sonia P. Juvik and James O. Juvik, 183–197. Honolulu: University of Hawai'i Press, 1998.

Silko, Leslie Marmom. "Landscape, History, and the Pueblo Imagination." In Robert Finch and John Elder, editors, *The Norton Book of Nature Writing*, 882–894. New York: W. W. Norton & Company, 1990.

Song, Jaymes. "Extinction Near with Native Bird's Death." Accessed October 15, 2011: http://archives.starbulletin.com/2004/12/01/news/story3.html.

Thompson, Nainoa. "Hawai'iloa." In *Wao Akua: Sacred Source of Life*, edited by Frank Stewart, 89–91. Honolulu: Division of Forestry and Wildlife, Department of Land and Natural Resources, State of Hawai'i, 2003.

Walker, George. 1999. "Geology." In *Manual of the Flowering Plants of Hawai'i*, revised ed., edited by Warren L. Wagner, Derral R. Herbst, and S. H. Sohmer, 21–35. Honolulu: University of Hawai'i Press and Bishop Museum Press, 1999.

"Way of the Dodo, The." Accessed November 5, 2011: http://www.loe.org/shows/shows.html?programID=08-P13-00031.

Woody, Todd. "Wildlife at Risk Face Long Line at U.S. Agency." *New York Times*, April 20, 2011. Accessed May 15, 2011: http://www.nytimes.com/2011/04/21/science/earth/21species.html?pagewanted=all.

Ziegler, Alan C. *Hawaiian Natural History, Ecology, and Evolution*. Honolulu: University of Hawai'i Press, 2002.

1: Journey to Hakalau

Baldwin, Bruce G., and Robert H. Robichaux. "Historical Biogeography and Ecology of the Hawaiian Silversword Alliance (Asteraceae): New Molecular Phylogenic Perspectives." In *Hawaiian Biogeography: Evolution on a*

Hot Spot Archipelago, edited by Warren L. Wagner and V. A. Funk, 259–287. Washington and London: Smithsonian Institution Press, 1995.

Carlquist, Sherwin. *Hawai'i: A Natural History.* Garden City, NY: Natural History Press for the American Museum of Natural History, 1970.

Carson, Hampton L. "Evolution." In *Atlas of Hawai'i,* 3rd ed., edited by Sonia P. Juvik and James O. Juvik, 107–110. Honolulu: University of Hawai'i Press, 1998.

Conant, Sheila. "Birds." In *Atlas of Hawai'i,* 3rd ed., edited by Sonia P. Juvik and James O. Juvik, 130–134. Honolulu: University of Hawai'i Press, 1998.

Drake, Donald R., and Dieter Mueller-Dombois. "Population Development of Rain Forest Trees on a Chronosequence of Hawaiian Lava Flows." *Ecology* 74 (1993): 1012–1019.

Gerrish, Grant, Dieter Mueller-Dombois, and Kent W. Bridges. "Nutrient Limitation and *Metrosideros* Forest Dieback in Hawai'i." *Ecology* 69 (1988): 723–727.

Givnish, Thomas J., Kendra C. Millam, Austin R. Mast, Thomas B. Paterson, Terra J. Theim, Andrew L. Hipp, Jillian M. Henss, James F. Smith, Kenneth R. Wood, and Kenneth J. Sytsma. "Origin, Adaptive Radiation and Diversification of the Hawaiian Lobeliads (Asterales: Campanulaceae)." *Proc. R. Soc. B* 276 (2009): 407–416.

Hart, Patrick J. "Tree Growth and Age in an Ancient Hawaiian Wet Forest: Vegetation Dynamics at Two Spatial Scales." *Journal of Tropical Ecology* 26 (2010): 1–11.

Juvik, James O. "Biogeography." In *Atlas of Hawai'i,* 3rd ed., edited by Sonia P. Juvik and James O. Juvik, 103–106. Honolulu: University of Hawai'i Press, 1998.

Lamoureux, Charles H. "Native Plants." In *Atlas of Hawai'i,* 3rd ed., edited by Sonia P. Juvik and James O. Juvik, 135–139. Honolulu: University of Hawai'i Press, 1998.

Lerner, Heather R. L., Matthias Meyer, Helen F. James, Michael Hofreiter, and Robert C. Fleischer. "Multilocus Resolution of Phylogeny and Timescale in the Extant Adaptive Radiation of Hawaiian Honeycreepers." *Current Biology* 21 (2011): 1838–1844.

Mueller-Dombois, Dieter. "Ohia Canopy Dieback a Natural Process." *Environment Hawai'i* 10:6 (December 1999).

Tuland, Mike. "The U.S. Department of Agriculture's Rural Development Approach to Alien Plant Control in Hawai'i: A Case Study." In *Alien Plant Invasions in Native Ecosystems of Hawai'i,* edited by Charles P. Stone, Clifford W. Smith, and J. Timothy Tunison, 577–583. Honolulu: University of Hawai'i Cooperative National Park Resources Studies Unit, 1992.

Wagner, Warren L., Derral R. Herbst, and S. H. Sohmer. *Manual of the Flowering Plants of Hawai'i,* revised ed. Honolulu: University of Hawai'i Press and Bishop Museum Press, 1999.

Ziegler, Alan C. *Hawaiian Natural History, Ecology, and Evolution.* Honolulu: University of Hawai'i Press, 2002.
Ziegler, Marjorie. "Wekiu." *Environment Hawai'i* 14:2, (August 2003): 4–5.

2: Place of Many Perches and Hooves

Cuddihy, Linda W., and Charles P. Stone. *Alteration of Native Hawaiian Vegetation.* Honolulu: University of Hawai'i Cooperative National Park Resources Studies Unit, 1990.
Dewar, Heather. "Forest Managers Seek to Stem the Tide of Loss with Recovery Projects Statewide." *Environment Hawai'i* 13:6 (December 2002): 1, 5–11.
Furukawa, George. "The Mother of the Rainforest." *American Forests* 107 (winter 2002): 40–43.
Hart, Patrick J. "Tree Growth and Age in an Ancient Hawaiian Wet Forest: Vegetation Dynamics at Two Spatial Scales." *Journal of Tropical Ecology* 26 (2010): 1–11.
Jeffrey, Jack, and Baron Horiuchi. "Tree Planting at Hakalau Forest National Wildlife Refuge." *Native Plants* (spring 2003): 30–31.
Jensen, Mari N. "Coming of Age at 100: Renewing the National Wildlife Refuge System." *BioScience* 53 (2003): 321–327.
Levy, Sharon. "Empty Nest Syndrome." *OnEarth* 27 (Summer 2005): 27–31.
Stone, Charles P. "Alien Animals in Hawai'i's Native Ecosystems: Toward Controlling the Adverse Effects of Introduced Vertebrates." In *Hawai'i's Terrestrial Ecosystems: Preservation and Management,* edited by Charles P. Stone and J. Michael Scott, 251–297. Honolulu: Cooperative National Park Resources Studies Unit, University of Hawai'i, 1985.
Stone, Charles P., and Linda W. Pratt. *Hawai'i's Plants and Animals.* Honolulu: Hawai'i Natural History Association, National Park Service, and University of Hawai'i Cooperative National Park Resources Studies Unit, 1994.
Tomonari-Tuggle, Myra Jean. *Bird Catchers and Bullock Hunters in the Upland Mauna Kea Forest: A Cultural Resource Overview of the Hakalau Forest National Wildlife Refuge, Island of Hawai'i.* Honolulu: International Archaeological Research Institute, Inc., 1996.
Tummons, Patricia. "At Hakalau Refuge, Hunter Pressure Overrides Conservations' Concerns." *Environment Hawai'i* 8:4 (October 1997): 1–3.
———. "Hawaiian Forests: How Do You Celebrate a Century of Loss?" *Environment Hawai'i* 13:5 (November 2002).
———. "Jack Jeffrey: Love for Birds Inspires His Art." *Environment Hawai'i* 13:1 (July 2002).
———. "Managers View as Mixed Blessing Proposed New Natural Area Reserves." *Environment Hawai'i* 11:9 (March 2001).
———. "Pasture or Trees? Graziers, Planters Pulled Territorial Government in Different Directions." *Environment Hawai'i* 13:4 (October 2002).

———. "The Roots of Ranching in Hawai'i: From Vancouver to Parker and Beyond." *Environment Hawai'i* 13:3 (September 2002).
US Fish and Wildlife Service. "Hakalau Forest National Wildlife Refuge." Last updated October 4, 2011. http://www.fws.gov/hakalauforest/.
Van Riper, Charles, III, and J. Michael Scott. "Limiting Factors Affecting Native Hawaiian Birds." In *Evolution, Ecology, Conservation, and Management of Hawaiian Birds: A Vanishing Avifauna*, edited by J. Michael Scott, Sheila Conant, and Charles Van Riper III, 221–233. Lawrence, KS: Cooper Ornithological Society, Allen Press Inc., 2001.
Ziegler, Alan C. *Hawaiian Natural History, Ecology, and Evolution*. Honolulu: University of Hawai'i Press, 2002.

3: Science to the Rescue?

Santiago, Louis S. "Use of Coarse Woody Debris by the Plant Community of a Hawaiian Montane Cloud Forest." *Biotropica* 32 (2000): 633–641.
Scowcroft, Paul. G. "Role of Decaying Logs and Other Organic Seedbeds in Natural Regeneration of Hawaiian Forest Species on Abandoned Montane Pasture." USDA Forest Service General Technical Report PSW-129 (1992): 67–73.

4: Laulima

Scowcroft, Paul G., and Jack Jeffrey. "Potential Significance of Frost, Topographic Relief, and *Acacia koa* Stands to Restoration of Mesic Hawaiian Forests on Abandoned Rangeland." *Forest Ecology and Management* 114 (1999): 447–458.
Scowcroft, Paul G., Frederick C. Meinzer, Guillermo Goldstein, Peter J. Melcher, and Jack Jeffrey. "Moderating Night Radiative Cooling Reduces Frost Damage to *Metrosideros polymorpha* Seedlings Used for Forest Restoration in Hawai'i." *Restoration Ecology* 8 (2000): 161–169.

5: Place of Many New Perches and Fewer Hooves

Wagner, Warren L., Derral R. Herbst, and S. H. Sohmer. *Manual of the Flowering Plants of Hawai'i*, revised ed. Honolulu: University of Hawai'i Press and Bishop Museum Press, 1999.
Ziegler, Alan C. *Hawaiian Natural History, Ecology, and Evolution*. Honolulu: University of Hawai'i Press, 2002.

6: Kill and Restore: Hawai'i Volcanoes National Park

Baker, James K., and Don W. Reeser. "Goat Management Problems in Hawai'i

Volcanoes National Park: A History, Analysis, and Management Plan." National Park Service Natural Resources Report No. 2. Washington, DC: US Department of the Interior, 1972.

Martin, Jim, and Laura Carter Schuster. "Case Study: Native Hawaiian Collection, Use, and Management of Plants and Plant Communities within Hawai'i Volcanoes National Park." *Ecological Restoration* 21 (2003): 307–310.

Reeser, Donald W. "Establishment of the Resources Management Division, Hawai'i Volcanoes National Park." In *Proceedings of the Seventh Conference on Research and Resource Management in Parks and on Public Lands,* 431–436. Hancock, MI: George Wright Society, 1993.

Taylor, D., and Larry Katahira. "Radio Telemetry as an Aid in Eradication of Remnant Feral Goats. *Wildlands Society Bulletin* 16 (1988): 197–199.

Tunison, Timothy J., and Chris Zimmer. "SEAs Successful as Alien Species Controls." *Park Science* 11:4: 1991.

US Department of the Interior. "Hawai'i Volcanoes." Accessed August 17, 2011. http://www.nps.gov/havo/index.htm.

7: The Pū'olē'olē Blows: Dry Forest Restoration at Auwahi, Maui

Medeiros, Arthur C. "The *Pū'olē'olē* Blows and 'Awa is Poured." In *Wao Akua: Sacred Source of Life,* edited by Frank Stewart, 113–119. Honolulu: Division of Forestry and Wildlife, Department of Land and Natural Resources, State of Hawai'i, 2003.

Medeiros, Arthur C., C. F. Davenport, and Chuck G. Chimera. "Auwahi: Ethnobotany of a Hawaiian Dryland Forest." Accessed October 30, 2010. http://www.hear.org/naturalareas/auwahi/ethnobotany_of_auwahi.pdf.

Medeiros, Arthur C., and Erica von Allmen. "Restoration of Native Hawaiian Dryland Forest at Auwahi, Maui." Accessed October 30, 2010. http://biology.usgs.gov/pierc/Pollution_%26_Ecological_Restoration/Dryland_restoration.pdf.

Rock, Joseph F. *The Indigenous Trees of the Hawaiian Islands.* Kaua'i and Rutland, VT: Reprinted by the Pacific Tropical Botanical Garden and Charles F. Tuttle, 1974.

Starr, Forest, and Kim Starr. "Natural Areas of Hawai'i." Last modified January 13, 2008. http://www.hear.org/naturalareas/auwahi/.

8: Turning Hands: Limahuli Botanical Garden, Kaua'i

Blaine, Jessica MacMurray. "Hawai'i's Limahuli Garden and Preserve." *Forest* (winter 2004): 39–41.

Fisher, Kathleen. "Limahuli Garden." *American Gardener* (July/August 1997): 38–39.

Maly, Kepā, and Onaona Maly. *"Hana Ka Lima, 'ai Ka Waha": A Collection of Historical Accounts and Oral History Interviews with Kama'āina Residents* and Fisher-People of Lands in the Halele'a-Nāpali Region on the Island of Kaua'i. Hilo, HI: Kumu Pono Associates, 2003.

McCormick, Kathleen. "Cultivating the Genuine Kauai." *New York Times,* May 26, 1996, 10-11, 15.

National Tropical Botanical Garden. "Limahuli Garden and Preserve." Accessed October 30, 2010. http://www.ntbg.org/gardens/limahuli.php.

9: **Multiple Perspectives**

Cabin, Robert J. *Intelligent Tinkering: Bridging the Gap between Science and Practice.* Washington, DC: Island Press, 2011.

10: **Nature Is Dead. Long Live Nature!**

Baldwin, A. Dwight, Jr., Judith De Luce, and Carl Pletsch, eds. *Beyond Preservation: Restoring and Inventing Landscapes.* Minneapolis: University of Minnesota Press, 1994.

Bawa, Kamaljit S., Nitin D. Rai, and Navjot S. Sodhi. "Rights, Governance, and Conservation of Biological Diversity." *Conservation Biology* 25:3 (February 2011), 639-641.

Burney, David A. *Back to the Future in the Caves of Kaua'i: A Scientist's Adventures in the Dark.* New Haven, CT: Yale University Press, 2010.

Cabin, Robert J. "Nature Is Dead. Long Live Nature!" *American Scientist* 101:1 (January-February 2013), 30-37.

———. "Science and Restoration under a Big, Demon Haunted Tent: Reply to Giardina et al. (2007)." *Restoration Ecology* 15:3 (September 2007): 377-381.

———. "Science-Driven Restoration: A Square Grid on a Round Earth?" *Restoration Ecology* 15:1 (March 2007): 1-7.

Caro, Tim, Jack Darwin, Tavis Forrester, Cynthia Ledoux-Bloom, and Caitlin Wells. "Conservation in the Anthropocene." *Conservation Biology* 26 (2012): 185-188.

Commoner, Barry. *The Closing Circle: Nature, Man, Technology.* New York: Knopf, 1971.

Cronon, William, ed. *Uncommon Ground: Rethinking the Human Place in Nature.* New York: W. W. Norton, 1996.

Davis, Clifton J., Ernest Yoshioka, and Dina Kafeler. "Biological Control of Lantana, Prickly Pear, and Hamakua Pamakani in Hawai'i: A Review and Update. In *Alien Plant Invasions in Native Ecosystems of Hawai'i,* edited by Charles P. Stone, Clifford W. Smith, and J. Timothy Tunison, 411-431. Honolulu: University of Hawai'i Cooperative National Park Resources Studies Unit, 1992.

Disneyland's Big Thunder Railroad Web site. Accessed October 15, 2011.

http://disneyland.disney.go.com/disneyland/big-thunder-mountain-railroad/?name=BigThunderMountainRailroadAttractionPage.
Disneyland's Redwood Creek Challenge Trail Web site. Accessed October 15, 2011. http://disneyland.disney.go.com/disneys-california-adventure/redwood-creek-challenge-trail/.
Donlan, Josh, Harry W. Greene, Joel Berger, Carl E. Bock, Jane H. Bock, David A. Burney, James A. Estes, Dave Foreman, Paul S. Martin, Gary W. Roemer, Felisa A. Smith, and Michael E. Soulé. "Re-wilding North America." *Nature* 436 (2005): 913–914.
Ehrlich, Paul R. *The Population Bomb.* New York: Ballantine Books, 1968.
Ellis, Jeffrey C. "On the Search for a Root Cause: Essentialist Tendencies in Environmental Discourse." In *Uncommon Ground: Rethinking the Human Place in Nature,* edited by William Cronon, 256–268. New York: W. W. Norton, 1996.
Feyerabend, Paul K. *Against Method: Outline of an Anarchistic Theory of Knowledge.* Atlantic Highlands, NJ: Humanities Press, 1974.
Freedman, David H. *Wrong: Why Experts Keep Failing Us—and How to Know When Not to Trust Them.* New York: Little, Brown and Company, 2010.
Giardina, Christian P., Creighton M. Litton, Jarrod M. Thaxton, Susan Cordell, Lisa J. Hadway, and Darren R. Sandquist. "Science Driven Restoration: A Candle in a Demon Haunted World—Response to Cabin (2007)." *Restoration Ecology* 15:2 (June 2007): 171–176.
Gobster, Paul H. "The Chicago Wilderness and Its Critics: III. The Other Side: A Survey of the Arguments." *Ecological Restoration* 15 (1997): 32–37.
———. "Restoring Nature: Human Actions, Interactions, and Reactions." In *Restoring Nature: Perspectives from the Social Sciences and Humanities,* edited by Paul Gobster and R. Bruce Hall, 1–19. Washington, DC: Island Press, 2000.
Gobster, Paul, and R. Bruce Hull, eds. *Restoring Nature: Perspectives from the Social Sciences and Humanities.* Washington, DC: Island Press, 2000.
Hardin, Garrett. "The Tragedy of the Commons." *Science* 162 (1968): 1243–1248.
Helford, Reid M. "Constructing Nature as Constructing Science: Expertise, Activist Science, and Public Conflict in the Chicago Wilderness." In *Restoring Nature: Perspectives from the Social Sciences and Humanities,* edited by Paul Gobster and R. Bruce Hall, 119–142. Washington, DC: Island Press, 2000.
Hemley, Robin. "To the Rainforest Room: In Search of Authenticity on Three Continents." *Orion* (May/June 2011): 38–45.
Higgs, Eric. "The Two-Culture Problem: Ecological Restoration and the Integration of Knowledge." *Restoration Ecology* 13 (2005): 159–164.
Hobbs, Richard J., Lauren M. Hallett, Paul R. Ehrlich, and Harold A. Mooney. "Intervention Ecology: Applying Ecological Science in the Twenty-First Century." *BioScience* 61:6 (2011): 442–450.

Ioannidis, John P. A. "Why Most Published Research Findings Are False." *PLoS Medicine* 2:8 (2005): e124.
Janzen, Daniel H. "Gardenification of Wildland Nature and the Human Footprint." *Science* 279 (1998): 1312–1313.
Jordan, William R., III. *The Sunflower Forest: Ecological Restoration and the New Communion with Nature*. Berkeley: University of California Press, 2003.
Kuhn, Thomas S. *The Structure of Scientific Revolutions*. 3rd ed. University of Chicago Press: 1996.
Lackey, Robert T. "Science, Scientists, and Policy Advocacy." *Conservation Biology* 21:1 (February 2007): 12–17.
Lehrer, Jonah. "The Truth Wears Off." *The New Yorker* (December 13, 2010): 52–57.
Mann, Charles C. "The Dawn of the Homogenocence." *Orion* (May/June 2011): 17–25.
———. *1493: Uncovering the New World Columbus Created*. New York: Knopf, 2011.
Marris, Emma. "The New Normal." *Conservation Magazine* (April–June 2010).
Martin, George P., Po-Yung Lai, and George Y. Funasaki. "Status of Biological Control of Weeds in Hawai'i and Implications for Managing Native Ecosystems." In *Alien Plant Invasions in Native Ecosystems of Hawai'i*, edited by Charles P. Stone, Clifford W. Smith, and J. Timothy Tunison, 466–482. Honolulu: University of Hawai'i Cooperative National Park Resources Studies Unit, 1992.
Martin, Paul S., and David A. Burney. "Bring Back the Elephants." *Wild Earth* (spring 1999): 57–64.
McKibben, Bill. *The End of Nature*. New York: Anchor Books, 1989.
Medeiros, Arthur C., Lloyd L. Loope, Patrick Conant, and Shannon McElvaney. "Status, Ecology, and Management of the Invasive Plant *Miconia calvescens* DC (Melastomataceae) in the Hawaiian Islands." Bishop Museum Occasional Papers 48 (1997): 23–36.
Menard, Louis. "Everybody's an Expert." The *New Yorker*. December 5, 2005: 98.
Meyer, Jean-Yves, and Jean-Pierre Malet. *Study and Management of the Alien Invasive Tree* Miconia calvescens *DC (Melastomataceae) in the Islands of Raiatea and Tahaa (Society Islands, French Polynesia): 1992–1996*. Technical Report 111. Honolulu: University of Hawai'i Cooperative National Park Resources Studies Unit, 1997.
National Tropical Botanical Garden's Web site. Accessed October 30, 2010. http://www.ntbg.org/.
Nelson, Richard K. *Make Prayers to the Raven: A Koyukon View of the Northern Forest*. Chicago: University of Chicago Press, 1983.
Nevle, Richard J., and Dennis K. Bird. "Effects of Syn-pandemic Fire Reduction and Reforestation in the Tropical Americas on Atmospheric Carbon Dioxide during European Conquest." *Palaeogeography, Palaeoclimatology, and Palaeoecology* 264 (2008): 25–38.

Omaha's Henry Doorly Zoo's Lied Jungle Factsheet. Accessed August 23, 2011. http://www.omahazoo.com/Post/sections/45/Files/Lied%20Jungle.pdf.

Palmer, Richard A. "Quasi-Replication and the Contract of Error: Lessons from Sex Ratios, Heritabilities and Fluctuating Asymmetry." *Annual Review of Ecology and Systematics* 31 (2000): 441–480.

Powell, Devin. "Columbus' Arrival Linked to Carbon Dioxide Drop." *Science News* 180:10 (2011): 12.

Ross, Laurel M. "The Chicago Wilderness and Its Critics: I. The Chicago Wilderness: A Coalition for Urban Conservation." *Ecological Restoration* 15 (1997): 17–24.

Sandbrook, Chris, Ivan R. Scales, Bhaskar Vira, and William M. Adams. "Value Plurality among Conservation Professionals." *Conservation Biology* 25:2 (April 2011): 285–294.

Shore, Debra. "The Chicago Wilderness and Its Critics: II. Controversy Erupts over Restoration in Chicago Area." *Ecological Restoration* 15 (1997): 25–31.

Turner, Frederick. "The Invented Landscape." In *Beyond Preservation: Restoring and Inventing Landscapes,* edited by A. Dwight Baldwin Jr., Judith de Luce, and Carl Pletsch, 35–66. Minneapolis: University of Minnesota Press, 1994.

Vince, Gaia. "Embracing Invasives." *Science* 331 (March 18, 2011): 1383–1384.

Wohlforth, Charles. "Conservation and Eugenics: The Environmental Movement's Dirty Secret." *Orion* (July/August 2010).

Ziegler, Alan C. *Hawaiian Natural History, Ecology, and Evolution.* Honolulu: University of Hawaiʻi Press, 2002.

Zimmermann, Helmuth G., V. Cliff Moran, and John H. Hoffmann. "The Renowned Cactus Moth, *Cactoblastis cactorum* (Lepidoptera: Pyralidae): Its Natural History and Threat to Native *Opuntia* Floras in Mexico and the United States of America." *Florida Entomologist* 84 (2001): 543–551.

Supplementary Sources for Additional Information

Apfelbaum, Steven I., and Alan W. Haney. *Restoring Ecological Health to Your Land.* Washington, DC: Island Press, 2010.

Berger, Andrew J. *Hawaiian Birdlife.* Honolulu: University of Hawaiʻi Press, 1981.

Cabin, Robert J. "Bird Survey Suggests If You Plant It, They Will Come." Accessed November 22, 2011. http://www.huffingtonpost.com/robert-j-cabin/bird-survey-suggests-if-y_b_849100.html.

———. "As Government Fails to Protect Endangered Species, One Grassroots Program Quietly Succeeds." Accessed July 10, 2011. http://www.huffingtonpost.com/robert-j-cabin/government-endangered-species_b_875425.html.

———. "Kill the Frogs?" Accessed November 20, 2011. http://www.huffingtonpost.com/robert-j-cabin/kill-the-frogs_b_853797.html.
———. "The Wildfires in Hawaii Are a Loss for Our World." Accessed September 18, 2011. http://www.huffingtonpost.com/robert-j-cabin/wildfires-in-hawaii-is-lo_b_840821.html.
Clewell, Andre F., and James Aronson. Forthcoming. *Ecological Restoration: Principles, Values, and Structure of an Emerging Profession.* 2nd ed. Washington, DC: Island Press.
Crosby, Alfred W. *The Columbian Exchange: Biological and Cultural Consequences of 1492.* Westport, CT: Praeger, 2003.
Diamond, Jared. *Collapse: How Societies Choose to Fail or Succeed.* New York: Viking, 2005.
———. *Guns, Germs, and Steel: The Fate of Human Societies.* New York: W. W. Norton, 1999.
Egan, Dave, Evan E. Hjerpe, and Jesse Abrams, eds. *Human Dimensions of Ecological Restoration: Integrating Science, Nature, and Culture.* Washington, DC: Island Press, 2011.
Friends of Hakalau Forest National Wildlife Refuge. Accessed November 15, 2011. http://www.friendsofhakalauforest.org/.
Hawaiian Ecosystems at Risk Project (HEAR). Last modified September 24, 2011. http://www.hear.org/.
"Hawai'i: Bird-Extinction Capital of the World." *Birder's World* 12 (April 2011).
"Hawai'i Conservation Alliance." Accessed on October 30, 2010. http://hawaiiconservation.org/.
Hawai'i Forest Industry Association. "Hawai'i's Dryland Forests: Can They Be Restored?" Accessed November 5, 2011. http://www.hawaiiforest.org/reports/dryland.html.
Jordan, William R., III, and George M. Lubick. *Making Nature Whole: A History of Ecological Restoration.* Washington, DC: Island Press, 2011.
Ka'ahahui O Ka Nāhelehele. "Nāhelehele Hawai'i Dryland Forest." Accessed October 15, 2010. http://www.drylandforest.org/.
Kingsbury, Noel. *Hybrid: The History and Science of Plant Breeding.* Chicago: University of Chicago Press, 2011.
Littschwager, David, and Susan Middleton. *Remains of a Rainbow: Rare Plants and Animals of Hawai'i.* Washington, DC: National Geographic, 2001.
Loope, Lloyd L. "Hawai'i and the Pacific Islands." In *Status and Trends of the Nation's Biological Resources,* edited by Michael J. Mac, Paul A. Opler, Catherine E. Puckett Haecker, and Peter D. Doran, 747–777. Reston, VA: US Department of the Interior, US Geological Survey, 1998.
Mehrhoff, Loyal A. "Endangered and Threatened Species." In *Atlas of Hawai'i,* 3rd ed., edited by Sonia P. Juvik and James O. Juvik, 150–153, Honolulu: University of Hawai'i Press, 1998.
Merlin, Mark D. *Hawaiian Forest Plants.* Honolulu: Pacific Guide Books, 1995.

Mulroney, Merryl J. *Treasures of the Rainforest.* Volcano, HI: Peregrine Fund, 1999.
Pacheco, Rob. "Inventory of a Koa." In *Wao Akua: Sacred Source of Life,* edited by Frank Stewart, 23–27. Honolulu: Division of Forestry and Wildlife, Department of Land and Natural Resources, State of Hawai'i, 2003.
Popper, Karl R. *The Logic of Scientific Discovery.* London, New York: Routledge, 2002.
Royte, Elizabeth. "On the Brink: Hawai'i's Vanishing Species." *National Geographic* 188 (1995): 2–37.
"Society for Ecological Restoration International." Accessed on October 30, 2010. http://www.ser.org/.
Stone, Charles P., and Linda W. Pratt. *Hawai'i's Plants and Animals.* Honolulu: Hawai'i Natural History Association, National Park Service, and University of Hawai'i Cooperative National Park Resources Studies Unit, 1994.
Stone, Charles P., Clifford W. Smith, and J. Timothy Tunison. *Alien Plant Invasions in Native Ecosystems of Hawai'i.* Honolulu: University of Hawai'i Cooperative National Park Resources Studies Unit, 1992.
Tetlock, Philip E. *Expert Political Judgment: How Good Is It? How Can We Know?* Princeton, NJ: Princeton University Press, 2006.
350.org. Accessed December 20, 2011. http://www.350.org/.
Tomich, P. Quentin. "Ohia: Adventures with a Genetic Marvel." In *Wao Akua: Sacred Source of Life,* edited by Frank Stewart, 31–37. Honolulu: Division of Forestry and Wildlife, Department of Land and Natural Resources, State of Hawai'i, 2003.
US Fish and Wildlife Service Endangered Species in the Pacific Islands Web site. Last Modified August 3, 2010. http://www.fws.gov/pacificislands/teslist.html.
Warshauer, F. R. "Alien Species and Threats to Native Ecology." In *Atlas of Hawai'i,* 3rd ed., edited by Sonia P. Juvik and James O. Juvik, 146–149. Honolulu: University of Hawai'i Press, 1998.
Wilkinson, Kim M., and Craig R. Elevitch. *Growing Koa: A Hawaiian Legacy Tree.* Holualoa, HI: Permanent Agriculture Resources, 2003.

INDEX

Page numbers in **boldface** type refer to illustrations

A

'a'ali'i (*Dodonaea viscosa*), 113–115, 119, 127
Acacia, 8–9, 189
Acacia koa: colonization and speciation, 8–9; decaying logs for regeneration, 42–44, 48, 68; as "forest engineer," 61–62; for gorse control, 16–18; growth rates, 21; harvesting, xxiii, 22; historical competition with 'ōhi'a, 114; multipodal stands, 77–78; old-growth stands, 20–21, 37–38; in planted corridors for restoration, 38, **39**, 55–56, **56**, 61, **62**, 63, 70–75, 77–78; as rain forest dominant, 4; using bulldozers to jumpstart regeneration of, 56–59
agriculture, naturalness of, 191–192, 198
ahupua'a, 78, 153, 157, 162
'akala (Hawaiian raspberry), 38, 51
'ākepa (*Loxops coccineus*), 74, 78
'aki ('akiapōlā'au, *Hemignathus munroi*), 74. See also Plate 7
'alalā (Hawaiian crow, *Corvus hawaiiensis*), 73, 175
alien plants, 86–87, 90–92, 94–95, 98–99, 102, 110, 132. See also *individual species*
Allerton Garden (National Tropical Botanical Garden), 136, 195
ālula (*Brighamia insignis*), 146–147

227

'amakihi (*Hemignathus virens*), 74, 115. See also Plate 8
animal rights lobby, 30, 33, 208
ants, xxv, 5, 99, 119, 170, 191
apapane (*Himatione sanguinea*), 28, 74
Auwahi, Maui. See dry forest restoration at 'Ulupalakua Ranch
'awa (*Piper methysticum*), 106–107
'awapuhi ginger (*Zingiber zerumbet*), 141

B

Baker, James, 85–88
balloon weed (milkweed, *Asclepias physocarpa*), 110, 118
banana *poka* (*Passiflora mollissima*), 34, 76
Bender, Dave, 141–143, 154
biological control, xxix, 15–16, 94, 199–201
birds: catching for feathers, 22, 148; colonization and speciation, 4–5; diseases, xviii, 27, 55; effects of goats on, 86; evolutionary and ecological history, 6, 197; flightless, xxii–xxiii, 6, 76, 148, 197; herbivory by, 76; importance for prehistoric navigation, xxiv; prehistoric abundance and extinction, xxiv, 21; use of tree corridors, 38, 55; using exotics to fulfill role of natives, 199. See also honeycreepers; *individual species*
Bishop Museum, 117, 136
bulldozers, 15–16, 41, 55, 118, 175
Burney, David, 159

C

Cactoblastis, 200–201
Canavalia kauensis, 88
canoes, xxii–xxiv, xxviii, 22, 29, 115, 131, 193, 202
Carlquist, Sherwin, 9
cats, 95
cattle: ecological impacts, 21, **23**, 24–25, 30–31, 40–41, 73, 102, 119, 137; importance of kikuyu to, 201; introductions of, xxv, 189; weed response to exclusion of, 34
Chicago Wilderness, 207–209
Clermontia lindseyana. See Plate 5
Clermontia pyrularia, 76
climate change, 55, 99, 187, 189–190, 197
Columbus, Christopher, 189, 190, 192
Commoner, Barry, 210
conch shell (*pū'olē'olē*), xxxii, 107
conservation: building support for, 32, 63–64, 124–127, 129; difficulty of engaging public in, 195–196; education and outreach, 63, 157–158, 179–181; and elitism, 186, 207–209; and eugenics, 206–207; history, 25–26; importance of partnerships, xix, 105–106, 109, 125–127, 178, 184; importance of pluralism, 176, 211–213; intellectual frameworks and challenges, xx, xxx–xxxi, 114, 172–176, 195–198; political opposition to, 25, 30–33, 85; role of science in, 180–186; spiritual dimensions, 104, 106–108, 118, 158; triage in, xxi, 177. See also ecological restoration
conservation community: attitudes towards academics, xix–xx; burnout, 68, 71; consensus on ungulate control, 26–27;

motivations, 123, 168–171; priorities, 177–180, 196, 199; purists and pragmatists, 172–176, 192–196; tensions within, xxvi–xxviii, 16, 26, 123, 167–168, 192, 196, 204, 211–212
Cook, James, xxiii, xxv, xxviii, 20–21, 29, 148, 189
Cyanea shipmanii, 76
Cytandra tintinnabula, 75

D

Dalton, Francis, 207
Darwin, Charles, 6, 207
decaying logs, in plant regeneration, 42–44, 48, 68. See also Nurse Log Experiment
Department of Hawaiian Home Lands, 14, 17
Department of Land and Natural Resources (DLNR), 94, 135, 138
disease, xviii, 10, 27, 55, 99, 191
Disneyland's nature adventures, 194–195, 198
Division of Forestry and Wildlife (DOFAW), 66, 103
dry forest restoration at 'Ulupalakua Ranch, Auwahi, Maui: botanical richness of region, 102; early conservation efforts, 103; ecological degradation, 102, **120**, **128**; endangered species, 103, 105, 109, 117, 128; fencing, 109, 118; first major outplanting, 106–107; introduction to story, xxxii; lessons learned, 122–124, 127–129; location, **xv** (map), 102; *māhoe* outplanting, 103–104, 116; management strategies, 106, 120–122, 124, 127–128; *Melicope adscendens* reproduction, 117; "natural" plant regeneration, 110, 114; partnerships, 105–106, 109, 125; Polynesian's use of flora, 115; refinement of techniques, 127–128; similarities to Big Island dry forest restoration project, 102, 105–106; site's advantages for restoration, 112; support from land owners, 105–106; volunteer involvement, 104, 106–109, **111**, 117–120, 122, 128

E

East Maui Watershed, 109, 125
ecological restoration: authenticity of, 193–196; diversity of people and institutions involved with, xxxi–xxxii; overarching management goals of, 22; paradigm of, xxxi, 193. See also conservation
education and outreach, 63, 157–158, 179–181
'elepaio (*Chasiempis sandwichensis*), 74
El Niño, 61–62, 103
endangered species, xxi, 47; at Auwahi, Maui, 103, 105, 109, 117, 128; at Hakalau Forest National Wildlife Refuge, 20–22, 26, 53, 62, 70, 74–77; at Hawai'i Volcanoes National Park, 94–95; at Limahuli Botanical Garden, 146, 160
Endangered Species Act, xvii, xix, 20
End of Nature, The (McKibben), 187
Erdman, Sumner, 106
Erhlich, Paul, 210
eugenics and racism, xxvi, 206–207

evolution, 4–9, 41, 46, 136, 170, 172, 174–175, 181
extinction, xvii, xxvii, xxxii, 20, 78, 170, 206; and goats, 86, 88; and Hakalau Forest National Wildlife Refuge, 21, 53; and Limahuli Botanical Garden, 160

F

fencing: as first step in restoration, 22; at Hakalau Forest National Wildlife Refuge, **23**, 31–34; at Hawai'i Volcanoes National Park, 87–88, **89**, 90, **93**, **98**; at Limahuli Botanical Garden, 137, 160; at 'Ulupalakua Ranch, Auwahi, Maui, 109, 118. *See also* Plate 13
fire, 35, 42, 72, 87, 91–92, 131; and alien species, 15, 34–35, 87, 102, 201; and Chicago Wilderness, 208; at Hakalau Forest National Wildlife Refuge, 21, 38, 42, 72; at Hawai'i Volcanoes National Park, 91–92; and indigenous people, xxvi, 207, 131; at Limahuli Botanical Gardens, 131. *See also* Plate 9
flies, Hawaiian pomace or vinegar (*Drosophila*), 6–7
Florida blackberry (*Rubus argutus*), 34, 42–43, 50
fountain grass (*Pennisetum setaceum*), 102, 110, 201
frost, 44, 56–59, 61–62

G

Galápagos, 196–197
Galápagos finches, 6
garden, as metaphor, 110, 127, 172–173, 191, 198

Garden of Eden, mythology of, xxii, xxxi, 188, 197
geese, 6, 75–76. *See also nēnē*
genetics, 7–8, 20, 29, 122, 127, 156, 173–175
ginger, 99–101; recovery after removal of, **100**
"Goat Management Problems in Hawaii Volcanoes National Park: A History, Analysis, and Management Plan" (Baker and Reeser), 85
goats, xxv, 15, 24, 27, 84–91, 97, 196. *See also* Hawai'i Volcanoes National Park: goat control
Godale, David, 153, 157
Goo, Don, 17–18, 39, 42, 48, 50–51, 60, 66–67, 69, 70
gorse (*Ulex europaeus*), 13, **14**, 15–19, 34–35, 68, 72, 75, 196. *See also* Plate 3
Grant, Madison, 207
"Green Cancer" (*Miconia calvescens*), 196
guava (*Psidium guajava*), 141, 147

H

Hakalau Forest National Wildlife Refuge: bird ecology, 72–75; creation of, 20; education and outreach from volunteer program, 63; effects of cattle removal, **23**, 34; endangered species, 20–22, 26, 53, 62, 70, 74–77; fencing, **23**, 31–34; "Friends of" organization, 71; frost protection of koa seedlings, 56–57, **58**, 59; hunter conflicts, 31–33; introduction to story, xxxii; jumpstarting koa regeneration with bulldozers, 41–42; koa corridors,

Index 231

38, **39**, 55, **56**, 61, **62**, 63, 72–73, 77–78; Kona subunit, 34; limitations of practical value of science, 54; location, **xv** (map), 3, 20; as microcosm of Hawaiian biodiversity and conservation, 21; noxious weeds, 34; Nurse Log Experiment, 43–45, 48, 50–51, 53, 67–68; plant propagation, 51–53, 59–61, 70, 75–76; Sport Hunting Plan, 32–33; ungulate control, **23**, 31–32; utilization of decaying logs for plant regeneration, 42–44; vegetation zones, 20–21; volunteer involvement, **56**, 61, 63–64, **65**, 66, 70–72. See also Plate 6
hala pepe (*Pleomele auwahiensis*), 110, 116
Haleakalā National Park, 84, 90, 103, 105, 109, 125
"*Hana ka lima, 'ai ka waha*" (Maly and Maly), 132
haole (foreigner), xvi, 127, 212
hāpu'u (tree fern, *Cibotium* spp.), 10, **12**, 28
Hardin, Garret, 210
hau (*Hibiscus tilliaceus*), 131, 135–136, 149–150
Hawai'i: characteristics of native biodiversity, 4–8; dispersal to and colonization of, 4–6; endangered species, xxi, 47; evolutionary history, 5–6; extinctions in, xvii, xxi–xxii, xxvi, 21, 53, 75–76, 86, 88, 99, 117, 198; flora of, 4, 46–47; geographic isolation, **xiv** (map), 5; geologic history, xxi; habitat loss and degradation, xviii, 21–23, 38; historic periods of, xxi–xxii; human population estimates, xxv; lack of basic ecological publications for, 41–42, 46; as microcosm, xx–xxi, 97; mysticism in botany and evolution, 46–47, 64; overview of archipelago, xxi; paradoxes of, xx–xxi; remaining native species and ecosystems, xviii–xix; settlement of, xxii–xxiv; speciation in, 4–6, 8, 12
Hawai'i (the Big Island), **xv** (map), 3, 5, 7, 20–21, 24, 30, 83, 95–97, 102–103, 201. *See also* Hakalau Forest National Wildlife Refuge; Hawai'i Volcanoes National Park
Hawai'i: A Natural History (Carlquist), 9
Hawaiian hawk (*'io*), 197
Hawaiian owl (*pueo*), 197
Hawaiian raspberry (*'akala*), 38, 51
Hawai'i creeper, 75
Hawai'i Forest Bird Survey, 20
Hawai'iloa, xxiii
Hawai'i Volcanoes National Park: alien plant control program, 91–94; conflicts with hunters, 85, 87; creation of, 83; creation of Resources Management Division, 89; dark-rumped petrel, 95; fencing, 87–90, **98**; goat control, 84–88, **89**, 90; hawksbill turtle, 95, **96**, 97; importance of endangered charismatic species, 94–95; introduction to story, xxxii; location of, **xv** (map); Mauna Loa silversword, **7**, 97; as microcosm of Hawaiian biodiversity and conservation, 97–99; *nēnē*,

95; perceptions of volcanic eruptions, 188; pig control, 90; Special Ecological Areas, 94
hawksbill sea turtle (*Eretmochelys imbricata*), 94–95, **96**, 97
Hillebrand, W. F., 76
Hilo, **xv** (map), 3–4, 14–15, 18, 59, 70, 94
Hispaniola, 189, 191
hoary bat, 5, 21–22
Homogenocene, 190
honeycreepers, xvii, 21, 73–75. *See also individual species*
Honolulu, **xv** (map), xxix, 26, 63, 117, 136
Horiuchi, Baron, 51, **52**, 53–54, 59–61, 64, 66, 70
horses, 189, 192
humans, relationship to nature, xx, xxii, xxviii–xxix, 178, 187–196, 198, 202, 208–209
hunters: alliance with animal rights lobby, 30; cooperation with conservationists, 109, 126; cultural differences from conservationists, 27, 29; effectiveness for controlling ungulates, 32–33; lobby for, 26–27, 31–32, 72, 85, 87
Hurricane Iniki, 134, 139

I

ihe (spear), 107
'*i'iwi* (*Vestiaria coccinea*), 39–40, 74, 180
indigenous peoples, xxv–xxvii, xxx–xxxi, 106, 189–191, 206–207, 212
indoor rain forest (Nebraska's "Lied Jungle"), 194
inventionist ecology, 193
'*io* (Hawaiian hawk), 197

J

Java plum (*Syzigium cumini*), 141
Jeffrey, Jack, 21–22, 34, 40–41, 54–59, 61–63, 65, 70–79, 201; awards, 71. *See also* Hakalau Forest National Wildlife Refuge
Judas Goats, 90

K

kamani (*Calophyllum inophyllum*), 66
kāma'o (*Myadestes myadestinu*), 78
kapa (cloth), 107
kapu (taboo), xxiv, xxvii, 132
Kaua'i, **xv** (map), xviii, xxv, 7, 78, 195–196
Kaua'i Endangered Seabird Recovery Project, 160
kauila, 107
kikuyu (*Pennisetum clandestinum*), 37–38, 72–73, 77, 102–103, 109–110, 113–114, 117–118, 127, 201
King Kamehameha, 24
koa. *See Acacia koa*

L

lama (*Diospyros sandwicensis*), 146
Leeward Haleakalā Watershed Restoration Partnership, 127
Leopold, Aldo, 41
Leopold Report, "Wildlife Management in the National Parks" (1963), 85
lichens, 8, 10, 12, 103, 116
Limahuli Botanical Garden: author's restoration experiment, 140–143; awards, 151, 159; Community-Based Subsistence Fishing Area designation, 162; cultural challenges, 153–154; endangered species, 146, 160; fencing, 137, 160;

Index 233

financial issues, 151–152, 159–160; fire-throwing ceremony, 131; history, 135–139; importance of education and outreach, 157–158; introduction to story, xxxii; location, **xv** (map); management strategies, 143–145, 155–156, 160, 162–163; ʻōʻopu recovery, 149–150; restoration programs, 134–135, 159–160, 162; Sierra Club volunteer project, 133, 145–147; stream, 134, 149; subdivisions, 133–134; volunteer involvement, 154–155. *See also* Plate 13
Little Ice Age, Europe, 190
lobelias, 41, 75–76. *See also* Plate 5
locals, importance of in conservation, xix, 32, 89, 108–109, 126–127, 152–154, 178–180, 183
loʻi (irrigated taro), 130, 132, **134**
Loope, Lloyd, 103

M
māhoe (*Alectryon macrococcus* var. *auwahiensis*), 103–104, 116
maile (*Alyxia oliviformis*), 116
Maly, Kepa and Onaona, 132
mamane (*Sophora chrysophylla*), 85
mammals, 5, 21, 27, 99, 197–198. *See also individual species*
Management Division at Hawaiʻi Volcanoes National Park, 89, 99
Manual of the Flowering Plants of Hawaiʻi (Wagner), 46–47, 117
Maui, **xv** (map), xviii, xxxii, 7, 18. *See also* dry forest restoration at ʻUlupalakua Ranch
Mauna Kea, **xv** (map), 4, 11–15, 20, 24, 37. *See also* Plate 2
Mauna Loa, **xv** (map), 4, 11–12, 95, 97

Mauna Loa silversword (Kaʻū silversword, *Argyroxiphium kauense*), 7, 94, 97, **98**
McKibben, Bill, 187, 189–190
meadow rice grass (*Ehrharta stipoides*), 37
Medeiros, Art, 103–110, 112–120, 122–125, 127–129, 201
Melicope adscendens, 117
Melicope knudsenii, 119
Merrill, Nancy, 133, 154, 157–158
Merwin, W. S., 129
mints, 22, 41, 75
modern period, xxii, xxv–xxviii, 202
mongooses, 86, 95, 189, 199–200
mosquitos, xviii, xxv, 27, 55, 132–133, 150, 189
moths, 12, 16, 116, 160, 200–201
mullein (*Verbascum thapsus*), 13
mycorrhizae, 44, 121

N
National Conservation Commission report, 206
National Tropical Botanical Garden, **xv** (map), xxviii, xxxii, 130, 135–137, 147, 153, 159, 161, 195
Native Americans, xxx–xxxi, 13, 189–190, 206
native flora: dispersal and colonization, 4–6; effects of ungulates on, 40–41, 85–88, 113; germination and propagation, 5, 42–44, 48, 50–53, 55, 72–73, 109–110, 122, 144–145; Plant Extinction Prevention Program, xix; speciation, 6–8; utilization by Polynesians, 115
Native Hawaiian Plant Society, 103, 106
Native Hawaiians. *See* Polynesians
Nature Conservancy, 103, 125

nēnē (*Branta sandvicensis*), 19–20, 37, 75–76, 94–95. See also Plate 4
New Nationalism speech, 206
noble savage, xxvi
Notch, Matthew, 133–135, 139–140, 143–151, 154, 156
novel ecosystems, 197–198
Nurse Log Experiment, 43–45, 47–50, 53, 67–68
nutrient cycling, 27, 91

O

Oʻahu, **xv** (map), 24, 59, 63, 123
octopus tree (umbrella tree, *Schefflera actinophylla*), 141
ʻōhiʻa (*Metrosideros polymorpha*): colonization and speciation, 5, 9–10; dieback, 10–11; growth rates, 21, 74; historical competition with koa, 114; mortality due to frost, 61; multipodal stands, 78; old-growth stands, 20–21, 37–38; polymorphism, 9–10; propagation, 51–52; as rain forest dominant, 4; use of decaying logs for regeneration, 42–44, 48. See also Plates 1 and 8
ola (life), 107
ʻōlapa (*Cheirodendron trigynum*), 10, 73
olomaʻo (Molokaʻi thrush, *Myadestes lanaiensis*), 78
ʻōmaʻo (Hawaiʻi thrush, *Myadestes obscurus*), 72–73
ʻōʻō (*Moho nobilis*), 78
ʻoʻopu, 149–150
orchids, 5, 194
ʻōʻū (*Psittirostra psittacea*), 21, 73, 78
outreach and education, 63, 157–158, 179–181

P

pāpala (*Charpentiera elliptica*), 131
pāpala kēpau (*Pisonia wagneriana*), 148
Parker Ranch, 196–197
Pele, 188
Perry, Lyman, 103–104, 116
petrel, dark-rumped (*Pterodroma phaeopygia sandwichensis*), 94–95
petrel, Hawaiian, 160
pigs: controversies about, 27, 29–30, 126; historical ecology, 29, 212–213; impacts, 26–27, **28**, 86, 95, 97, 119; recovery following removal, **23**, 90
Pinchot, Gifford, 206–207
Piper methysticum (ʻawa), 106–107
Pipturus kauaiensis, 142
Planck, Max, 212
Plant Extinction Prevention Program, xix
Pōhaku-o-Kāne, 131, 133
Poliʻahu, 12
pollination, 5, 8, 116–117, 122, 127, 147, 192, 198
Polynesians: bird-catching, 148; cessation of round-trip sailing voyages, xxiii–xxiv; conceptions of time, 132; contemporary perspectives on prehistoric extinctions and pig ecology, xxvii–xxviii, 210, 212–213; culture, xxiv, 131–132, 140; discovery of and settlement in Hawaiʻi, xxii–xxiv; ecological effects, xxiii–xxvi, 22–23; fire-throwing ceremony, 131; Mauna Kea Observatory conflict, 12; in National Conservation Commission report, 206; pig management and hunting,

29–30; population estimates of, xxv; species introduced by, xxiii, 129, 141, 149; stonemasons, 155; taro cultivation, 130–132
Polynesian Voyaging Society, xxiii
poʻolā (*Claoxylon sandwicense*), 115
poʻouli, xvii–xix, xxiv
Population Bomb, The (Ehrlich), 210
population growth, as environmental problem, 210
practitioners, lack of respect for, 16–17, 182–183
predation: control efforts, xviii, xxiii, 6, 20, 75, 95, 97, 162; escape from, 16, 26, 86, 191; on seeds, 45, 50
prehistoric period, xxi–xxii, xxiv–xxix, 8, 76, 132, 202
prehuman period, xvi, xxi–xxiii, xxix, 44, 197, 202
prickly pear, 200–201
pristine ecosystems: emphasis on, 179, 184, 192; false vision of, xxii, xxvi, xxx; loss of in Galápagos, 197; rain forest, 41, 52, 126, 139
pueo (Hawaiian owl), 197
pūʻōlēʻolē (conch shell), xxxii, 107

R

racism and eugenics, xxvi, 206–207
rain forest ecology, 4, 10–11, 40–41, 44, 77–78
ranchers, 13–15, 24–26, 85, 102, 125, 196–197, 201
Reeser, Don, 84–90
rewilding, 198
Rock, Joseph, 102, 104
rodents, xxiii, xxv, 86, 95, 104, 122, 127, 132, 178–179, 189, 197, 199
Roosevelt, Teddy, 206–207

S

Saddle Road, 3–4, 8, 13, 70–71
science: author's first research experience, 45; limitations of, 54, 124, 142–143, 155–156, 202–207; and politics, 206–207; practical relevance of, 48, 142–143, 182–186, 203–204; problems of basing policies on, 206–209; publications, 45–46; short-lived nature of theories and "truths" in, 205–206, 209–210, 212–213; value of pluralism in, 210–213
sea turtle eggs, 29
seed banks, 15, 41–42, 72–73, 91, 144
seed dispersal: by birds, xxviii, 55, 73; by humans, 173, 207; by nonnative species to replace extinct native species, 198; by ungulates, 22, 27, 30, 213
seeds: consumption, 6, 12, 16, 45, 50, 104; germination and propagation, 5, 42–44, 48, 50–53, 55, 72–73, 109–110, 122, 144–145; scarification, 41–42
sheep, 13, 15, 24, 27, 97, 189
Sierra Club, 71, 133, 139, 145
Silva, Rene, 124
silverswords, 7, 85, 94, 97, **98**
snails: giant African, 199; native, xvii, xxix, 6, 115, 162, 199; rosy cannibal, 199
Somers, Phyllis, 133, 147, 151, 153, 158
spear (*ihe*), 107
St. John, Harold, 136–137
sweet vernal grass (*Anthoxanthum odoratum*), 37

T

Taino Indians, 189, 191
taro, xxv, xxxiii, 130–132, **134**, 149
Thoreau, Henry David, xxx

Tragedy of the Commons, The (Hardin), 210
tree corridors, 37–39, 55–56, 61–63, 72–75, 77–78
tree snails, xxvii, xxix, 199
trial-and-error, 35, 47, 54–56, 59–61, 142
Tummons, Pat, 25–26

U

uluhe (*Dicranopteris linearis*), 10, 77
'Ulupalakua Ranch, 104–105, 118. *See also* dry forest restoration at 'Ulupalakua Ranch
ungulates: as cause of ecological degradation, 22, **23**, **28**, 40–41, 85–88, 92, 97–98, 109, 147, 199; conflicts over, 27, 29–33, 85; control of, 22, 27, 31–32, 91, 125, 160, 199; effects on water supply, 25, 125; and plant ecology, 35, 103, 113, 147; recovery following removal of, **23**. *See also individual species*
University of Hawai'i, 30, 59, 94, 106, 138, 160
Urakami, Alan, 17–18, 39, 42, 48–51, 67–70
US Fish and Wildlife Service, xviii, 20, 31, 33–34, 59, 66, 105
US Forest Service, 3, 15–17, 40, 41–42, 51, 66–68, 84, 94, 139, 141, 196
US National Park Service, xxxii, 30, 59, 84–85, 88–89, 93, 96, 98, 100

V

Vancouver, George, xxv, 24
VanderWerf, Eric, xviii

velvet grass (*Holcus lanatus*), 37
Volcano Village, 3, 83
volunteers. *See* dry forest restoration at 'Ulupalakua Ranch: volunteer involvement; Hakalau Forest National Wildlife Refuge: volunteer involvement; Limahuli Botanical Garden: volunteer involvement
von Allmen, Erica, 111–113, 120–123, 127

W

Wagner, Warren, 46–47
Wass, Richard, 32–33
watershed partnerships, 125–127
water supply, 24–26, 125–126, 170
wēkiu bug (*Nysius wekiuicola*), 12
Wichman, Charles "Chipper", 135–139, 141–142, 147–151, 153, 155–156, 158–159, 162–163
Wichman, Haleakahauoli, 136, 138–139
Wichman, Juliet Rice, 135–137, 139, 150, 153, 158, 163
wilderness: critique of concept, xxxi, 178, 188; as entertainment, 194–195; and indigenous peoples, xxx–xxxi; perspectives on, xxviii, xxix–xxxi, 27, 63, 112, 188, 200
Wilderness Act, xxxi
"Wildlife Management in the National Parks" (The Leopold Report, 1963), 85
Winter, Kawika, 159–160, 162

Y

yellow jackets, 99
Yoshioka, Joan, xix

ABOUT THE AUTHOR

ROBERT J. CABIN is associate professor of ecology and environmental science at Brevard College. Before returning to academia, he worked as a restoration ecologist in Hawai'i for the U.S. Forest Service and the National Tropical Botanical Garden. He is the author of *Intelligent Tinkering: Bridging the Gap between Science and Practice* (2011).

PRODUCTION NOTES FOR...

Cabin / *Restoring Paradise*
Cover design by Julie Matsuo-Chun
Interior design and composition by Julie Matsuo-Chun
  with display type in ScalaScans and text in Minion Pro
Printing and binding by Thomson-Shore, Inc.
Printed on 60# Joy, 434 ppi